A Skywatcher's Year

Have you ever wondered what that bright point of light twinkling near the horizon is, or just when you can expect to see the best shooting stars of the year? *A Skywatcher's Year* has answers to these and other questions about what is visible in the night sky throughout the year. Through 52 essays, *A Skywatcher's Year* guides readers to celestial events and phenomena that occur or are visible with the naked eye and binoculars for each week of the year. It acquaints readers not only with up-to-date astronomical informa- tion on stars, nebulae, meteors, the Milky Way, and galaxies, but also conveys the beauty and wonder of the night sky. Covering both the Northern Hemisphere and the Southern Hemisphere, *A Skywatcher's Year* helps readers find prominent stars and constellations, bright star clusters, nebulae, and galaxies, and explains how and when to observe prominent annual meteor showers.

Jeff Kanipe began his career as a science journalist, writing scripts for the astronomy radio program Star Date, which was heard on radio stations across the United States and Canada. He has many years of writing experience, first with *Astronomy* and more recently with *Sky & Telescope*. Based in Dallas, Texas, Jeff Kanipe is a prolific writer of articles and features on astronomy.

A Skywatcher's Year

Jeff Kanipe

CAMBRIDGE
UNIVERSITY PRESS

PUBLISHED BY THE PRESS SYNDICATE OF THE UNIVERSITY OF CAMBRIDGE
The Pitt Building, Trumpington Street, Cambridge, United Kingdom

CAMBRIDGE UNIVERSITY PRESS
The Edinburgh Building, Cambridge CB2 2RU, UK   http://www.cup.cam.ac.uk
40 West 20th Street, New York, NY 10011–4211, USA   http://www.cup.org
10 Stamford Road, Oakleigh, Melbourne 3166, Australia

First published 1999

Printed in the United Kingdom at the University Press, Cambridge

Typeset in Utopia 9.5/13.5 pt. in QuarkXPress® [SE]

A catalogue record for this book is available from the British Library

Library of Congress Cataloguing in Publication data

Kanipe, Jeff, 1953–
    A Skywatcher's Year / by Jeff Kanipe.
        p.   cm.
    ISBN 0 521 63405 9 (pb)
    1. Astronomy – Popular works. 2. Astronomy – Observer's manuals.
    3. Stars – Observer's Manuals. I. Title.
    QB44.2.K35   1999
    520–dc21   98-41631 CIP

ISBN 0 521 63405 9 paperback

For my mother and father, who were always my pole stars,
and for my two daughters, Hayley and Carly, who will one day be pole
stars themselves.

Contents

Foreword

The stars are presented to you here as you'll come to know them in the heavens, one by one and in easily identified patterns, as they appear with satisfying orderliness in the course of a single year. The earliest sky-watchers must have come to know the stars in just this way, constellation after constellation, with each celestial grouping identified with the sights, smells and sounds of the changing seasons.

Today, we know the stars move across the sky's dome each night because Earth spins on its axis once a day. We know that new stars continually appear in the east because our vantage point on the Galaxy shifts, as we orbit the Sun once a year. Skywatchers now have the power to see the stars – with ordinary binoculars, or sophisticated telescopes and photographic equipment – in ways that would have astounded our ancestors. Our cosmology has taken us out of the center of everything, and made us inconceivably small in contrast with the rest of the universe. We live in a universe of mind-boggling collisions, explosions, and energies. And yet, in a way that binds us to the first skywatchers, the stars still seem a part of us, and we a part of them.

Even in our modern world, where lights of the cities often obscure the stars, it seems that nearly everyone has a 'gene' for astronomy. Thus to those of us already acquainted with astronomy, there is only one pleasure greater than skywatching, and that is to encourage someone else to look up! It's a pleasure to commend this book to you, and to imagine your going outside with it tucked under your arm, to begin your own exploration of the boundless sky.

Deborah Byrd
Earth & Sky Radio Series

Acknowledgments

Many thanks to Robert Burnham and Richard Berry for their advice and wisdom and to my good friend and artist Tim Jones, who constantly reminds me that 'fun' is the point of it all.

I am grateful to Alexandra Witze, who not only read and copy-edited the manuscript in its rough form, but who also supported and encouraged me during the writing phase, which, as usual, went on too long. *Siempre tu, Alejandra.*

Jeff Kanipe

Author's Note

This book is based on a series of weekly columns I wrote in 1992 and 1993 during my editorship of *StarDate* magazine, which is still published by the University of Texas' McDonald Observatory. The column, called 'Skywatch,' was distributed to about 20 or 30 newspapers in the United States by the Associated Press, but found its greatest readership by far on the Internet. I had no idea how popular 'Skywatch' was until I announced one day that I was going to stop writing the column so I could devote more time to my editing duties. In less than three days, I received over 500 emails from around the world imploring me to reconsider. I did, at least for another year, when, exhausted, I threw in the towel. I avoided logging on for a month.

In their original form, the 'Skywatch' columns were brief essays describing some interesting sky phenomenon or event for a particular week in a particular year. In this new incarnation, *A Skywatcher's Year* may be considered an outdoor program guide of natural sky events that are visible every year. In this respect, it differs from astronomical almanacs, which emphasize one-time events happening in a specific year for specific locales. I've listed some astronomical almanacs in a list of Further Reading at the end of this book, any one of which would make an excellent companion to *A Skywatcher's Year*.

Another difference between this book and the column is that I've endeavored to describe celestial phenomena visible not only in the Northern Hemisphere but in the Southern Hemisphere as well. The observability of the sky in both hemispheres, however, is latitude dependent. Hence, people living near the equator are unlikely to see aurorae or noctilucent clouds, while those in the high northerly latitudes will have a poor view, if at all, in the direction of the center of the Milky Way. Nevertheless, in each essay I've tried to offer something for everyone no matter what part of the world they live in. The night sky truly is a moveable feast of wonders, both by season and by latitude.

Introduction

Familiarizing yourself with the night sky

Like our ancestors did thousands of years ago, it's quite natural for us today to view the night sky as an inverted bowl upon which the stars are attached and the Sun, Moon, and planets move. And it requires only a little imagination to further visualize that the inverted bowl of stars is actually the hemisphere of a globe – a celestial sphere – which we see from the inside out. Earth floats freely in space within this sphere, which lies at an immense but arbitrary distance. The sphere revolves slowly around our world, incrementally conveying the constellations from east to west until they return to their original positions a year later.

Of course, we recognize that the celestial sphere is an illusion: the Moon and Sun lie at very different distances, the stars are light-years away, and it is Earth's rotation and orbit that make the sky appear to move. But for purposes of tracking the motions and directions of celestial bodies, as well as learning the stars and constellations, a two-dimensional celestial sphere is an effective mental contrivance.

If we consider the sky as a sphere rotating about an axis, then it must possess directional bearings, a north and a south pole, and an equator. In the case of the celestial sphere, these are essentially projections into space of the Earth's cardinal directions, poles, and equator.

A compass can help you locate north, south, east, and west from your observing location, although it can be done just as well by noting where the Sun rises and sets (especially around the equinoxes). Although it is not important to be rigorously precise in noting the cardinal directions, it helps to know them in a general sense for skywatching purposes. When I describe a star or constellation rising in the 'northeast,' or lying 'due south,' and you know where these directions are, then you have won half the battle of getting yourself oriented.

Dividing the sphere, and hence the sky, into east and west halves is a circle called the celestial meridian. The meridian extends across the sky over your head and through the north and south points of your horizon. (There is also a circle crossing the horizon through the east and west points called the 'prime vertical,' but this doesn't come into play in this book as often as the meridian.) A celestial body is said to 'culminate' when it crosses the meridian. Hence, when I say that a certain star 'lies on the meridian (or culminates) tonight at 9 o'clock,' you know to look for it on a north-to-south line at the star's highest point in the sky.

Notice that I said 'the *star's* highest point.' All stars reach different altitudes above the horizon that are dependent on your latitude. In the Northern Hemisphere, stars toward the south never get very high in the sky, while those in the north arch overhead, or nearly so. Southern stars do, of course, rise higher the further south you go, but then stars in the northern sky begin to slip lower. Again, this adds to the illusion that the sky is one half of a sphere.

On the other hand, there is indeed a point in the sky that can be called 'highest,' with respect to you, the observer. It always lies directly over your head and is called the zenith. As such, the zenith is 90° (i.e., at a right angle), from all points on your horizon and also lies on the meridian.

The north celestial pole lies very near the end of the handle of the Little Dipper and is marked by the star Polaris (more popularly known as the North Star). In the Northern Hemisphere, its angular distance above your horizon is equivalent to your latitude. So, if you live at latitude 35° N, then Polaris lies 35° above your northern horizon. The closer you get to the North Pole, the higher the star appears in the sky, until at the North Pole, it stands at your zenith.

Conversely, the south celestial pole in the Southern Hemisphere is in the constellation Octans and marked by the faint star Sigma (σ). Its distance above any horizon is also latitude dependent. Since the entire celestial sphere pivots around these two points in the sky, they never rise or set. (For more on finding the celestial poles, see 'Finding true north, March 22 – 28,' and 'Finding true south, March 29 – April 4.')

By definition, the celestial equator is the great circle lying halfway between the north and south celestial poles. It is Earth's equator projected into space. Hence, if you were standing on the equator, the celestial equator would be the prime vertical circle running overhead through the east and west points on the horizon. In the Northern Hemisphere, the celestial equator lies south of your zenith at an angular distance equivalent to your latitude. Thus, if you live at latitude 30°, the celestial equator lies south of your zenith by that amount. (You can also say that at latitude 30° the celestial equator is inclined to the south horizon by 60°.)

The final great circle to note is the ecliptic, which passes through the twelve constellations of the zodiac, plus a thirteenth, Ophiuchus. It is, in essence, the projection of the mean plane of Earth's orbit into space. Since Earth's orbital plane is coincident with that of the solar system, the annual paths of the Sun, Moon, and planets lie very near the ecliptic circle. And because Earth's axis is tipped over nearly 23.3° with respect to this plane, the ecliptic, too, is inclined to the celestial equator by that amount.

Four equidistant points on the ecliptic represent solar milestones, when the Sun either crosses the celestial equator or is at its greatest distance from it. These points represent the four seasonal transitions. The spring and autumnal equinoxes occur where the ecliptic intersects the celestial equator. In spring, the Sun crosses the celestial equator moving northward; in autumn, it is heading south. The summer solstice is the point where the Sun reaches its northernmost extreme from the celestial equator; the winter solstice is the Sun's southernmost reach. Again, because the Earth's axis is tilted 23.3°, this is how much the Sun moves above and below the celestial equator.

Directions in the sky

When finding your way around in the city, you know that to go 'north' or 'south' means to head toward the north or south cardinal point on the horizon. However, when a celestial object or event is said to be visible in the 'north,' you should look in the sky in a direction toward the north celestial pole. A star that is said to lie ten degrees 'south' of another star means that you look ten degrees from that star in a direction that leads toward the south celestial pole. 'Northeast' in the sky, then, would fall between cardinal east and the north celestial pole, and 'southwest' would be between cardinal west and the south celestial pole. If a celestial pole is not visible from a particular hemisphere (as the south celestial pole is not for those in the Northern Hemisphere, and vice versa), then its location must be inferred from the region of sky that lies directly opposite the visible celestial pole.

Sometimes, as when you are facing the southern sky in the Northern Hemisphere, the directions for east and west will seem reversed. After some practice, these directions will come more naturally.

A matter of degrees

One of the best ways to find a faint star or a deep-sky object, such as a galaxy or star cluster, is to look for it in relation to a bright, well-known star. But when you look at the bright star, which is easy to see, how do you tell someone how far away the fainter object is from it? You could say that object X is five 'inches' north of star Y, but that assumes that everyone agrees on how big an inch appears on the sky. Such a method of judging apparent distance is not only awkward but inaccurate.

Astronomers measure apparent distances and separations on the sky using the angular scale: degrees, minutes of arc, and seconds of arc. In this book, we will be mostly concerned with degrees. In the above example, then, we might say that object X is '5°' north of star Y. Fine. But how much is a degree?

Everyone knows that a circle contains 360°, and a half circle 180°. A circle traced across the celestial dome (essentially half a sphere) from one point on the horizon to the point opposite would equal 180°. From the horizon to the zenith would be half that amount, or 90°. Half that distance again, midway up into the sky from the horizon, is 45°; half again is 22.5°, and so on.

This gives us an approximate scale that can be applied to quadrants of sky, but what about smaller areas in and around constellations? We need something familiar to help us gauge smaller chunks of sky. That 'something' is your fist. A fist at arm's length covers about 10° of sky; 12° if your hands are large. (For more on calibrating fist size, see 'Spring arrives,' March 15 – 21.) From horizon to zenith, you should be able to mentally stack nine fists. An even finer gauge is the Moon. The Moon's apparent diameter is about half a degree (as is the Sun's). So two full Moons side by side equal one degree.

We can go to a still smaller scale. Consider that a single degree consists of 60 minutes of arc, and each minute of arc consists of 60 seconds of arc. If the Moon's apparent diameter is half a degree, then it can also be described as being 30 minutes of arc, or 1,800 seconds of arc, in apparent diameter. (We say 'apparent' diameter as opposed to 'true' diameter to describe the size of these bodies as they appear on the celestial sphere. Obviously the actual sizes of the Moon, Sun, and planets are much greater than their angular sizes.)

The planets, too, exhibit tiny disks in the sky that are measured in seconds of arc. When nearest Earth, Mars has an apparent diameter of 25 seconds of arc; Jupiter, 50 seconds of arc; and Venus nearly a minute of arc.

Star magnitudes

In addition to their myriad numbers, stars also come in a wide range of brightnesses. It is very often much easier to find a particular star or star group if you can determine beforehand how bright it is. To do so, we need some standard scale of brightnesses to use as a gauge.

In the second century B.C., the ancient Greek astronomer Hipparcos became the first to develop a qualitative scale for determining the brightness of a star. In the Hipparcos system, naked-eye stars were sorted into six different classes of brightnesses, or magnitudes. Twenty of the brightest stars were given first-magnitude status. The faintest stars that could be discerned were designated magnitude 6. Stars that fell between these two extremes were given intermediate magnitudes – 2, 3, 4, and 5.

With the advent of the telescope, however, countless stars fainter than magnitude 6 were observed and their brightnesses had to be measured more precisely. By the middle of the nineteenth century, astronomers agreed to refine the magnitude scheme in order to make it a quantitative rather than a qualitative scale. A difference of a single magnitude now corresponds to a brightness ratio of 2.5. Hence, a star of magnitude 1 is 2.5 times brighter than a star of magnitude 2, and 100 times brighter than a star of magnitude 6.

The original 20 first-magnitude stars ranged so widely in brightness that the very brightest, by virtue of the revised magnitude system, were assigned negative and zero values. In order of diminishing brightness, these are: Sirius, -1.46; Canopus, -0.72; Arcturus, -0.04; Rigel Kentaurus (Alpha Centauri), 0.00; Vega, 0.03; Capella, 0.08; Rigel, 0.12; Procyon, 0.38; Achernar, 0.46; Betelgeuse, 0.50; Hadar, 0.61; Altair, 0.77; Aldebaran, 0.85; and Antares, 0.96. The star Spica, at 0.98, is almost exactly a magnitude 1 star.

If it shines or reflects light in the sky, it has an apparent magnitude, not only the planets, the Moon, and the Sun, but meteors and artificial satellites as well. Venus, at its brightest, is magnitude -4.7; the Moon is magnitude -13; and the Sun is magnitude -26.5. On this scale, the Sun is about 9 billion times brighter in light output than Sirius, which is 25 magnitudes fainter. Sirius, on the other hand, is about a trillion times brighter than the faintest star (magnitude 28.5) visible with the Hubble Space Telescope.

Occasionally, I'll describe objects that can best be seen in binoculars or a small telescope. Whereas the naked eye can see only to magnitude 6 or so, a good pair of binoculars in dark skies can see stars as faint as magnitude 10; a 4-inch telescope, magnitude 12; an 8-inch, magnitude 14. Even if objects can be glimpsed with the naked eye, like the Andromeda Galaxy or the Orion Nebula, more features can be seen in those objects when you are able to gather more of their light. More light translates into more detail or, to invoke a technical term, greater 'resolution.' That's where even slight optical aid can help enhance the beauty of objects in the night sky.

Stellar nomenclature

At least fifty of the brightest stars have names that are either of Greek or Latin origin (Sirius, Castor, Pollux), or Arabic (Altair, Fomalhaut). Many of the brighter naked-eye stars in the constellations, however, are designated not by name but by small letters of the Greek alphabet. In this system, introduced in 1603 by Johannes Bayer, the brightest star in a constellation generally is labeled Alpha or α, the second-brightest star is Beta or β, and so on down to Omega or ω. (There are important exceptions to this, however, as I will soon describe.)

The name of a star in the Bayer system is the Greek letter designation followed by the possessive of the constellation's Latin name. And, in fact, this is how astronomers still refer to these stars. For example, the brightest star in the constellation Canis Major, besides being called Sirius, is Alpha (α) Canis Majoris. (However, astronomers typically refer to the proper name of a star of that magnitude.) In the essays, I use both the Greek symbol as well as the name of the Greek letter to aid in helping you find the star on a star map.

As mentioned, the Bayer system, in a general sense, goes in order of descending brightness. But not always. For example, the seven brightest stars in Ursa Major, which comprise the Big Dipper, are lettered in order of their position from the cup, because they are not much different in brightness from one another. Thus, the brightest star in Ursa Major is fifth down, Epsilon (ε) Ursa Majoris, rather than Alpha. There are other exceptions as well. In Orion, Rigel is quantitatively the brightest star in that constellation, but Betelgeuse, the second-brightest, is designated Alpha, while Rigel is designated Beta. The same is true in Gemini, in which Pollux, the constellation's brightest star, is designated Beta, while Castor, the second-brightest star, is designated Alpha.

For now, these things are all you need to know to get started observing the night sky. The few remaining terms relating to distances, types of stars, and other astronomical nomenclature will be defined in the text. You don't need to be versed in astronomical jargon to enjoy the night sky purely from an aesthetic point of view, but just as knowing where each of the cardinal directions lie from your home or how to read a road map can help you find your way from one place to another, these modest terms will help you move effortlessly from star to star until you know the heavens as well as your own neighborhood.

PART I

The Sky by Seasons

You do not have to sit outside in the dark. If, however, you want to look at the stars, you will find that darkness is necessary. But the stars neither require nor demand it.

Annie Dillard

Winter

Winter, which usually arrives on or about December 22, never seems to be appropriately on time, no matter where you live. In the southern realms of the Northern Hemisphere, the first day of winter can be quite mild, even tropical, while in the north, cold air has been infiltrating since late September and several inches of snow may already be on the ground. Depending on where you live, you are either surprised that winter has snuck up on you or incensed that someone would make a big deal about the season's entrance long after its arrival.

Those of us living in the Northern Hemisphere may well take heart in the first day of winter, for it means that the Sun has reached its most southerly extreme – it's summertime in the Southern Hemisphere – and will soon be heading back toward the north, bringing with it more direct light and longer days. In fact, by late January, and definitely by mid-February, you can tell that it doesn't get dark quite as early as it did in mid-November and December.

Even as the Sun halts its southerly advance and begins its slow return to the north, the coldest days still lie ahead for those in the Northern Hemisphere. Granted, with daylight saving time not in effect during the autumn and winter months, the days seem that much more abbreviated, but still by December 22 the Sun sets around 5 o'clock and it is dark by 6 o'clock. (The Sun sets even earlier the further north you are.) Arctic air whisks in from the north whipping up little whirlwinds of leaves or snow, while over-running warm air from the south covers the sky with an unbroken blanket of slate-gray stratus clouds. For northerners, the winter solstice is not cause for celebrating the Sun's turnaround; it just means that the weather will be getting colder before it starts getting warmer again.

Why is this the case? With each passing day, the Sun's rays shine more directly on the ground. Shouldn't it, therefore, be coldest at the winter solstice and warmer thereafter?

It would, if Earth's surface temperature were solely dependent on the changing angle of the Sun. But it's not. Day-to-day weather is more dependent on the balance of incoming solar heat and outgoing radiation reflected from both the surface and the atmosphere. As long as more heat is being reflected than absorbed, the temperature falls. Hence, in the Northern Hemisphere around December 21 or 22, even though the Sun's elevation at noon is lowest and the daily duration of sunshine is at its minimum, the coldest days occur in January and February, as the Sun advances northward. Not until the rate of heating overtakes the rate of radiative cooling do the daily temperatures begin to rise, as they do in the spring.

Conversely, after the first day of summer, with the Sun slipping ever southward, temperatures reach their warmest weeks later, in July and early August. They don't begin to abate until the amount of heat the Earth returns to space exceeds the amount it receives, which occurs during the autumn months. This delay between the official onset of winter and summer and their associated temperatures is called the lag of the seasons.

For some latitudinal extremes in the Northern Hemisphere, the lag of the seasons can be ludicrous. Very often, cooler weather doesn't become noticeable in the extreme south until well into October, and even then, the few cool days are adjoined by a string of warm, humid ones. For example, in south Florida and Texas (latitude 26° N to 30° N or so), mild Christmases are not unusual, and you can often sit in your shirtsleeves sipping iced tea while basking in the warm sun in January. Conversely, I've seen nearly ten inches of heavy, wet snow dumped on Milwaukee (latitude 42° N) in April (almost a month after the spring equinox), and near-freezing temperatures in early June, (a couple of weeks before the summer solstice).

In such cases there doesn't seem to be so much of a seasonal lag as a seasonal identity complex, which, as I said earlier, often leads to surprise and malediction. The weather hasn't caught up with the month it's supposed to occur in, and we end up hurling invectives at the calendar.

Imagine for a moment, though, what life would be like if the Sun never performed its annual shift across the sky, but was always directly overhead at local noon at the equator? Then, as seen from everywhere else, the Sun's path would transcribe the same daily arc across the sky, rising and setting at the same point on the horizon and at pretty much the same time every day. There would be no change in the seasons and little or no variation in weather. Those who lived in the northern and southern extremes of the world would forever experience cool or cold weather, while those living closer to the equator would be subjected to invarying heat.

Fortunately, Earth's axial tilt of 23.3° saves us from such celestial ennui.* Over weeks and months, the Sun appears to move gradually north, then south, then back again. Seasons change, stabilize, destabilize, then change again. Weather fluctuates, winds shift, sunlight and shadows vary. The sky transforms itself as the clouds and stars of the seasons come and go. And whether we acknowledge it or not – or give in to it – these rhythms are embedded in our culture in many ways. We may hunt, gather, or fall in love beneath a Harvest Moon, brew up deep discussions around cozy winter fireplaces; let our spirits soar like well-struck baseballs in the clear skies of a temperate spring; and become like children at play again in the summer.

Even if humans did not exist on this planet, Earth itself would still be compelled to reflect the changes wrought throughout the year by the simple, inherent tilt of its axis. Since we are here to experience these changes, we can't help but embody them, too, personifying not only the seasons, but our very connection to the stars.

Also this week:
- The Andromeda Galaxy is overhead in the evening sky for observers in the Northern Hemisphere. (See November 1 – 7.)
- Bright star Achernar (magnitude 0.4) in Eridanus the River is near the meridian around 7 o'clock in the Southern Hemisphere and for observers in the more southerly latitudes of the Northern Hemisphere.

---

* More accurately, the tilt of Earth's axis amounts to 23°.26'. I've rounded up to 23.3° because, for general purposes, 4 arcminutes is not that significant.

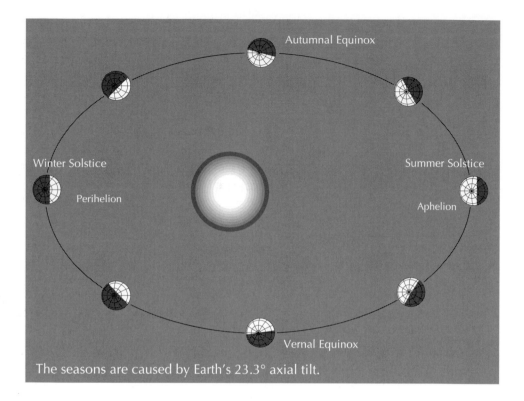

The seasons are caused by Earth's 23.3° axial tilt.

December 28 – January 3
New year, new millennium, same stars

Although much ado is made about the changing of centuries on Earth, the stars hardly notice. The ones visible in the evening sky this week are no different than those present when the nineteenth century gave way to the twentieth. Moreover, the brightnesses and patterns of the stars tonight are virtually indistinguishable from those seen shining several millennia ago in the skies of Earth.

Had you been on ship's deck with Columbus this week in 1493, just prior to his return voyage across the Atlantic from Cuba, or even if you had accompanied Aristotle in 325 B.C. as he took an evening stroll in the Lyceum in Athens, the stars would have looked no different than they do now.

If we could go back this week to Aristotle's time and look up at the night sky, we would still find wasp-waisted Orion the Hunter, with his distinctive 'belt' of three stars aligned east to west, well up in the northeastern sky by 8 o'clock. Following closely behind would be the sparkling Dog Star, Sirius, the brightest star in the night sky. Nearly overhead, the tiny cluster known as the Pleiades, or Seven Sisters, would glimmer like diamond chips just above the triangular visage of Taurus the Bull, his angry eye marked by the ruddy star Aldebaran. To the northeast, another bright star, Capella in Auriga the Charioteer, would flash its usual yellow and orange hues. Nearer the eastern horizon, two stars side by side would be just clearing the trees: orange Pollux, the brightest and easternmost of the two, and white Castor. These are the two brightest stars in Gemini the Twins.

The winter Milky Way with its clouds of stars would stretch across the top of the sky, just as it does now, from southeast to northwest. The giraffe-shaped constellation Perseus the Hero, the zig-zag grouping of Cassiopeia the Queen, and the crude house-shaped pattern of Cepheus the King would be holding court in the upper vault of the sky in the north. Dominating the western quadrant of the sky would be the Great Square of Pegasus with long, lithe Andromeda the Maiden trailing after in the form of an elongated V-shaped strand of stars.

If you want to see a noticeably different sky, you'll have to time travel at least ninety thousand years into the past or the future, when the true motions of the nearby stars through space would cause more dramatic changes in the shapes of some of the constellations. As we will see, the stars do in fact move, but their motion is only apparent on astronomical, not terrestrial, scales. Furthermore, stars eventually burn out and die, but only over periods of hundreds of millions, even billions, of years. Compared to our brief lifespans on Earth, the stars are truly eternal.

The year's first meteor shower

There is one aspect of the sky visible this week that you wouldn't have seen five, four, or even three hundred years ago: a meteor shower.

In the Northern Hemisphere, there are eight major meteor showers each year, 'major' being defined as ones that produce at least 20 meteors per hour during their peak. The Quadrantids, which peak each year between January 3 and 4, is one of those. First reported during the nineteenth century, this brief, one-night display is testimony to how circumstances can change in our tiny corner of the universe.

The Quadrantid shower is one of the more unpredictable meteoric displays. It has been known to produce as few as a dozen to as many as 250 meteors an hour. The usual number cited is 95, although it is doubtful you will see that many, unless you are observing from a dark location during the shower's short-lived peak. During the latter part of the twentieth century, some veteran meteor watchers believed the shower was intensifying. More observations will be needed, however, to confirm this or not. What is certain is that you never know what you're going to get with the Quadrantids, so perhaps it's worth a look.

The shower gets its name from the now retired constellation Quadrans Muralis (a mariner's quadrant depicted in a mural), which lay in the region between Ursa Major and Boötes (closer to the latter). Today, the point in the sky from which the meteors appear to emanate is located in Boötes.

In early January, this conical-shaped constellation rises in the northeast around midnight and is well up by 2 a.m. You can't miss its most prominent member, Arcturus, which is distinctly reddish and twinkles violently when near the horizon. Arcturus is the second-brightest star in the northern night sky. To observe the shower, go outside after midnight – say between 2 and 3 a.m. – and face northeast. Fairly dark skies are necessary because the Quadrantids are known to be fast and faint with a bluish color to their trails. Although it goes without saying, bundle up against the cold. Sleeping bags or a reclining chair help increase your comfort level, as does the proverbial cup of hot cocoa.

In deciding if you want to get up that early in the morning or not, you would do well to consider the old meteor-watcher's proverb: 'You may see a few, you may see many, but stay in your bed and you won't see any.'

Also this week:
- Fomalhaut sets in the southwest soon after the Sun. (See September 13 – 19.)
- Variable star Mira in Cetus crosses the meridian around 7 o'clock.

January 4–10
The flashing Dog Star

On a winter's evening, with the snow crunching beneath your boots and your breath hanging in frosty clouds in the air, your eye may catch upon a single star glimmering like a multifaceted diamond through the dark silhouettes of bare trees in the southeast. As you gaze upon this scene, you may imagine a kind of stillness settling upon the world, or perhaps within yourself. At such contemplative times it is hard to tell. But something about 'lone stars' strikes an emotional chord within us all. Can there be no greater symbol of solitude and yet dignified defiance than a single star shouting down the night?

Perhaps no star in our hemisphere evokes such feelings more than Sirius the 'Dog Star,' which is the star you see in the southeastern sky during the early evenings this week. Sirius is not only the brightest star in the constellation Canis Major (the Greater Dog), it is also the brightest star in the night sky. The only other star seen from Earth that is brighter than Sirius is the Sun. But though the Sun may outshine Sirius from our perspective, if you could place the two side-by-side in the sky an equal distance away, Sirius would be the brighter star – 23 times brighter, in fact. The reason: Sirius is not only nearly twice as large as the Sun, but it is a much hotter type of star and hence more intrinsically luminous.

Although Sirius may look like a single star, it isn't alone in space. Orbiting Sirius is a small companion star, which was found in 1862 by Alvan George Clark, an eccentric but venerated lens maker, who, at the time of the discovery, was testing the optics of one of his telescopes for a Chicago observatory. (I'd say the test was a success.) Unfortunately, this companion, called Sirius B, is not readily apparent in small telescopes; it takes a powerful, well-tuned set of optics to catch its glint next to Sirius A, which is 10,000 times brighter.

Nevertheless, a window of opportunity to glimpse this tiny sun is opening for observers with larger refracting and reflecting telescopes. Remember that Sirius B orbits Sirius A in an elliptical orbit, and as such there are times when it lies farther from the blinding star than others. Sirius B completes one orbit around Sirius A every 50 years. In 1994, the little star reached a minimum separation distance of about 3 arcseconds. Since then, this distance has been increasing. During the first 25 years of the twenty-first century, the separation distance between the two stars will grow to a maximum separation (in 2025) of 11.5 arcseconds. By then a 4-inch refractor or 6- to 8-inch reflector should be able to split the pair.

Until then, you're going to need access to a high quality 12-inch reflector or 6- to 7-inch refractor to see Sirius B. (If your local astronomy club or college observatory conducts a public star party or starwatch during the winter months, you can ask them to try and split Sirius A and B for you.) Look southeast of Sirius for the companion shining in the brighter star's glare. By 2002, Sirius B should be about 5.5 arcseconds from Sirius A and should just be visible in telescopes of this size and quality. The task becomes easier in 2005, when the separation widens to 7 arcseconds, and in 2010, the separation is about 9 arcseconds, with the companion due east of Sirius.

Things just keep getting better until 2025 when the star reaches maximum separation, called apastron. (The star will then be northeast of Sirius.) Thereafter, the separation decreases until the next minimum, called periastron, in 2043.

Astronomers classify the companion as a white dwarf, the core of a dying star that has collapsed so much under its own gravity that it is very dense. This companion has roughly twice the diameter of Earth, but its mass nearly equals the Sun's: a cupful would weigh over 50 tons.

Sirius' intense brightness, and the fact that it doesn't rise very high in the sky as seen from the Northern Hemisphere, conspire to trick people unfamiliar with the stars into believing they are witnessing a UFO. Whenever it is seen near the horizon, be it in the east or west, the star's brightness with respect to the landscape endows Sirius with the illusion of nearness. Furthermore, the agitated mixing of warm and cool air cells in the atmosphere bends the rays of starlight in random directions like a prism, making Sirius appear to shift rapidly in color from red to blue as well as jiggle around in position. The rapidly changing colors and the apparent jiggling animate the star to such an extent that it seems to be a nearby UFO maneuvering anomalously in the sky. Some terrified individuals have even reported being 'pursued' by Sirius, a phenomenon not unlike the illusion of being followed by the Moon.

The Dog Star is indeed nearby, though not in terrestrial terms. At 8.6 light-years, or a little over 51 trillion miles, Sirius is the fifth-nearest star to the Sun. Astronomically speaking, this is a mere hop, skip, and a jump away. Nonetheless, it would take tens of thousands of years to cross this distance using conventional spacecraft. Astronomical velocities are required to traverse astronomical distances.

So, when you look at Sirius this week, consider it a neighbor – like a farm on the far horizon with the house lights on. Truth is, no star that we can see this time of year is so far from Earth that its light can't warm our spirits on an otherwise cold winter night.

January 11–17
The Milky Way in winter

Most everyone knows about the summer Milky Way, the beautiful, coarse band of stars that arches across the sky from the south to north-east horizons in July and August. (See June 28 – July 4; July 5 – 11; July 12 – 18.) The winter sky, too, has its Milky Way, though it differs from the summer Milky Way in that it is slightly fainter and more difficult to see from areas near light-polluted cities.

Still, if you are fortunate enough to live in an area free of light pollution, or if you are able to get away to such a location, you can see the winter Milky Way this time of year high overhead around 9 o'clock. Like its summer counterpart, the winter Milky Way is punctuated by many bright star clusters, some of which, in a dark sky, look like tiny, isolated clouds to the unaided eye. But whereas the brightest clusters in the summer Milky Way appear predominantly in its southern reaches – in the constellations Sagittarius and Scorpius – the winter star clusters are distributed fairly evenly throughout, giving the winter Milky Way an overall softer luster. Moreover, the summer Milky Way is bisected by what look like dark rifts, which are actually clouds of interstellar dust blocking the light of more distant background stars. The winter Milky Way, too, contains such dust, but it is not as obvious.

Why do the winter and summer Milky Ways look different? It largely has to do with Earth's vantage point in our home galaxy. If we could transport ourselves outside our galaxy, we would see that it is really a flat disk of hundreds of billions of stars interspersed with dark lanes of interstellar dust. The Sun is embedded near the mid-plane of this disk in a quiet galactic 'outback' located some 27,000 light-years from the center of the Milky Way.

On summer evenings, the Earth's night side faces toward the center of our galaxy. Like the heart of a great metropolis, this area contains the densest and brightest star clouds. In winter, however, our nightscape is in the opposite direction, toward the outer arms of the Galaxy, where the stars thin out like porch lights on the edge of town. Hence the winter Milky Way presents a more demure vista.

The Milky Way actually traces out our galaxy's spiral arms, which wind out of its center like a pinwheel. The arms contain the dust and gas that give rise to stars. The stars, which at first seem so randomly placed, actually help astronomers map our galaxy's structure from the inside out. Astronomers have mapped out two specific spiral arms in the winter sky: the Orion Arm, which the Sun inhabits, and the Perseus Arm, which is located at a greater distance along the galactic plane. The summer section of the Milky Way extending into Sagittarius is called, predictably enough, the Sagittarius arm.

Although the winter Milky Way may not appear as glorious as the summer Milky Way, it contains more bright stars in a single region than its complement. With the Milky Way high in the sky, as it is between 9 and 10 o'clock in mid-January, run your eye from the zenith toward the south-southeastern horizon. In this quadrant alone, you should be able to count eight very bright stars – nine if you are located far south enough to see Canopus in the constellation Carina. This is the heart of the winter Milky Way.

January 18–24
The winter clock

Many people live under the mistaken belief that they cannot learn to recognize constellations. They may recall looking up on some clear summer night in their past and feeling overwhelmed by the thousands of points of light that seem scattered randomly across the sky. But the stars are, in fact, quite easy to learn if you use their constant positions as pointers to other stars and star groups.

In the Northern Hemisphere, if you go outside this week around 9

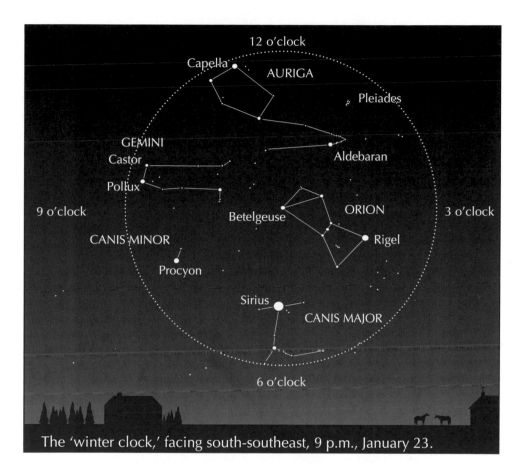

The 'winter clock,' facing south-southeast, 9 p.m., January 23.

o'clock and face south-southeast, you will see the hard-to-miss constellation Orion the Hunter. You know you're looking at Orion when you see three stars of the same brightness lying next to each other in a distinctive row stretching southeast to northwest. This line of stars represents the hunter's belt. The belt itself is framed within four stars, two of exceptional brightness that are diagonally opposite one another. The smoldering red star at the upper left of the belt, marking one of the Hunter's shoulders, is Betelgeuse.

If you stand back and take in this grand canvas of sky, you'll notice that red Betelgeuse lies near the center of an oval ring of bright winter stars that occupies the entire southeastern quadrant of the sky. This huge asterism is sometimes referred to as the 'winter circle' or 'winter hexagon,' but it can just as easily be considered a kind of giant starry clock, with Betelgeuse being the clock's centerpoint and the region's brightest stars consigned to certain hours.

We'll assign the 6 o'clock position to the brightest star visible in the entire night sky, dazzling white Sirius in Canis Major. Orion's belt conveniently points to Sirius, though it is so bright it is difficult to misidentify with other stars. At the 8 o'clock position, then, lies Procyon, a bright creamy yellow star in Canis Minor (the Lesser Dog). About 4° to the northwest is Gomeisa, or Beta (β) Canis Minoris, a magnitude 3 star. A simple line between these two stars is all that makes up the pattern of this humble constellation.

Snugged between the 9 and 10 o'clock positions, respectively, are two stars of almost equal brightness: Pollux and Castor. These stars mark the heads of Gemini the Twins. The northernmost star, Castor, is white, while Pollux is orange.

At nearly the 12 o'clock position, and thus high overhead, is Capella, a yellow star located in the pentagonal constellation Auriga the Charioteer. Dropping down to about 1 o'clock we see red Aldebaran, the bright, angry eye in the triangular face of Taurus the Bull. Completing the circuit is blue-white Rigel, at 4 o'clock.

Do you see any other fixed 'stars' as bright as these shining above Orion in the swath of sky between 8 and 1 o'clock? If so, you're seeing Mars, Jupiter, or Saturn. These bright planets move along the ecliptic path, which runs north of Orion through Gemini and Taurus, and often appear as 'guests' in the winter circle.

As the night progresses and these constellations shift to the west, the clock face lists a bit – forcing you to tilt your head a little to orient yourself to the clock face. Just remember that Sirius always marks the six o'clock position. That way, the winter clock can make an excellent reference with which to familiarize yourself with this area of the night sky, no matter what time you look at it.

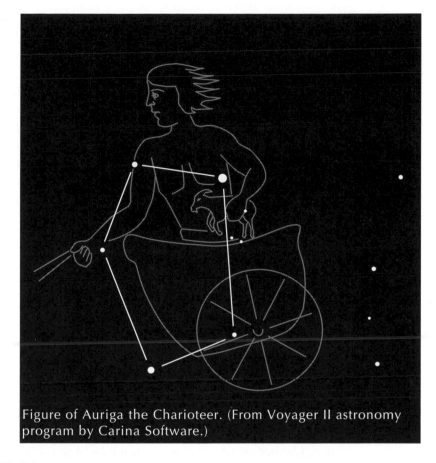

Figure of Auriga the Charioteer. (From Voyager II astronomy program by Carina Software.)

January 25–31
Who was Auriga the Charioteer?

With the winter Milky Way mounting high in the sky in late January evenings, it is a good time to locate a bright constellation that represents both a star lore mystery and high mythological drama. I'm speaking of Auriga the Charioteer, a pentagonal figure that resembles either a squatty roofed house or a crude baseball home plate, whichever form you prefer.

For people at mid-northern latitudes (30–40° N), Auriga is first seen early December evenings rising just after sunset, following first the zig-zag pattern of Cassiopeia then the giraffe-shaped Perseus into the sky.

By early January, Auriga is almost straight overhead at 8 o'clock. Just walk outside and look up. In the Southern Hemisphere at latitude 33° S, Auriga stradles the north meridian this week around 9 o'clock, but, even then, it has a median altitude of only 22°.

The most prominent stellar member in Auriga is a brilliant yellow-white star on the northwest side of the pentagonal pattern. This is Capella, third-brightest star in the northern sky, sixth-brightest in all the sky, and the nearest of the brightest stars to the northern pole star. Capella's name means 'Goat Star,' and indeed, the Charioteer is usually depicted in ancient star maps clutching a goat in his arms. Just south of Capella, the goat motif is continued by a distinctive triangular group of stars called 'the kids.'

But who is this charioteer and why is he tending goats? The second question was addressed by star lore expert Julius D. W. Staal in his book *The New Patterns in the Sky*. Staal tells us that ancient charioteers, in addition to driving the chariots of their rulers, also watched over the animals kept in the royal stalls. Capella, he writes, may represent Amalthea, a female goat said to have sustained the young Jupiter with her milk during his stay on the island of Crete, where Ops, his mother, had hidden him from his cruel father, Saturn.

Another possibility is suggested by Richard Hinkley Allen, author of *Star Names, Their Lore and Meaning*. He too notes that Capella in some star tales represents the goat Amalthea. But Allen also cites an earlier version in which Amalthea is personified by a nymph who feeds the infant Jove goat milk and honey on Mount Ida. And in still another inter- pretation, also mentioned by Allen, Capella represents neither nymph nor goat, but a goat's horn broken off by young Jupiter while romping playfully with the animal. In the heavens, the horn becomes the Cornucopia, or the Horn of Plenty.

So much for the identity of the goat star. Now who or what is the cha- rioteer? Donald Menzel, reviser of Martha Evans Martin's classic work, *The Friendly Stars*, proposes that Auriga may delineate the chariot driven by Neptune or Poseidon, god of the sea, pulled by a team of sea horses. The beauty of this explanation is that it brings to the celestial stage this deity of the sea who, though not normally represented in the famous mythological star tale 'The rescue of Andromeda' (see October 11 – 17), plays a catalytic role.

Several other charioteers in ancient literature are said to represent Auriga, but the two most popular are Erichthonius and Myrtilos. Erichthonius, son of Vulcan and Minerva, had inherited his father's lameness. To acquire greater mobility, Erichthonius devised the chariot. Zeus was so impressed with the boy's ingenuity that he placed the inventor and his invention in the sky.

The saga of Myrtilos, however, is a rather sordid tale of betrayal and hapless love worthy of a Shakespearean drama. In short, Myrtilos, son of Hermes, was a charioteer who betrayed his master King Oenomaus of Pisa on behalf of the king's daughter, the beautiful Hippodamia.

The tragedy was set in motion when word reached King Oenomaus that an oracle had predicted Hippodamia's husband would be responsible for the king's untimely death. Hippodamia, however, had no husband because she had yet to marry. To prevent her from doing so, King Oenomaus decreed that anyone vying for Hippodamia's hand in marriage must first compete with him in a chariot race. Since Oenamaus had only the best-bred horses, which were kept in prime condition by Myrtilos, his victory, and Hippodamia's marital status, would remain foregone conclusions. (The losers, by the way, were put to death.)

One day, Pelops, who claimed to be the son of Tantalus, king of Lydia, offered to compete for Hippodamia's hand in the deadly chariot race. Hippodamia immediately fell in love with this handsome youth, but she knew what the outcome of the race would be unless she took matters into her own hands. Knowing of Myrtilos' infatuation for her, Hippodamia deluded him into thinking that she would love him if he did her one little favor: sabotage the king's chariot. Blind with love, Myrtilos readily did as she asked. He removed the brass lynch pins in the king's chariot and replaced them with wax. During the race, the king was dragged violently to the ground by his own horses, and in his dying breath cursed Myrtilos.

With the king now out of the way, Hippodamia and Pelops were free to marry, which they did in Myrtilos' presence. Feeling the sting of betrayal, the star-crossed Myrtilos later tried to rape Hippodamia while Pelops was away, but he was caught in the act by the enraged husband, who threw Myrtilos to his death into the sea. Pelops ultimately became ruler of all Pisa.

And, I suppose, they lived happily ever after.

As intriguing as these star myths are, when I see Auriga and Capella this time of year, I prefer to think of the constellation as a sign that warmer weather is on the way. (Indeed, Capella was considered by some ancient cultures to be a harbinger of spring.) While winter's grip is still firmly in place, it's reassuring to look just south of the constellation, beneath the wheels of the chariot in Taurus, and know that, come early June at midday, the Sun will be there.

Also this week:
- Regulus, brightest star in the constellation Leo the Lion, rises by 8 o'clock. (See April 12 – 18.)

February 1–7
The second-brightest star in the night sky

One of nature's more pleasing coincidences is that the two brightest stars in the night sky – Sirius (mag − 1.46) and Canopus (mag − 0.72) – lie only 35° from one another – that's about three and a half fist-widths held at arm's length. What's not so pleasing, however, is that for many sky-watchers in the Northern Hemisphere Canopus never rises very high in the southern sky. Its peak altitude at latitude 36° N is only about 1°. To see Canopus, you're going to have to travel south.

In Dallas (latitude 32.4° N), the circumstances are somewhat more favorable. Canopus appears in the south-southeastern sky around 7:30 p.m. this week and, over the next four hours, transcribes a 30° arc from rise to set. Still, it never gets more than four and a half degrees above the horizon when it transits, which, this time of year, is around 9:30 p.m. local time.

But what a difference a few hundred miles make. In places like Corpus Christi, Texas (latitude 27.4° N), and Tampa, Florida (latitude 27.5° N), Canopus transcribes a 51° arc and is in the sky for over six hours. At its highest point, it reaches an altitude of nine and a half degrees. Places even further south – Miami, the Hawaiian Islands, and locales in Mexico, northern India, and Africa – fare even better.

Because of its southern location, Canopus doesn't have the popular status in the Northern Hemisphere of, say, Vega or Capella, which occupy the high latitudes. Consequently, you don't often see much written in the popular vein about Canopus, despite its being the second-brightest star in the night sky. Not so in the Southern Hemisphere, where Canopus is circumpolar at south latitudes greater than 35° S.

According to star-name authority and etymologist Richard Hinkley Allen, Canopus represents the helmsman of the ship *Argo Navis*, the vessel that carried Jason and his Argonauts in their mythic search for the golden fleece, and which was once a constellation unto itself. Argo Navis originally occupied a vast irregular area of sky some 45 by 75 degrees, but was later subdivided by astronomers into Puppis the Stern, Vela the Sail, and Carina the Keel, in which Canopus marks the ship's rudder. Appropriately enough, this 'pilot' star is used in modern times as a navigational reference by astronauts and interplanetary spacecraft.

Though Canopus lies nearly nine times further away than Sirius (75 light-years), it is nonetheless an eye-catching winter beacon. To appear as bright as it does, Canopus must be intrinsically brighter than Sirius. Actually, Sirius is the hotter star, but Canopus is some 15 times larger – 30 times larger than the Sun. A large surface area radiates more light than one that is smaller, though hotter. If Canopus were at the same distance as Sirius, it would easily be the brightest object in the sky, with the exceptions of the Sun and Moon.

In the southern part of the Northern Hemisphere this time of year (south of latitude 32° N), Sirius rises almost two hours before Canopus does, so you'll need to look below Sirius and slightly west. The star is white, though sometimes it appears gold tinged, due to the fact that, being low in the sky, its light rays refract through more atmospheric layers to reach our eyes than a star with greater altitude. This golden color has led some star lore experts to theorize that the name Canopus is really a derivation of the Egyptian *Kahi Nub*, which means 'golden earth.'

Anyone who has stood on a beach or the deck of a cruise ship while in the more southerly realms of the Northern Hemisphere and beheld Sirius and Canopus together in the sky may get the sense of seeing neighboring suns. In fact, with the exception of Betelgeuse and Rigel, all of the bright stars of the winter circle (see January 18 – 24) can be considered the Sun's neighbors: Sirius, 8.6 light-years; Procyon, 11 light-years; Pollux, 35 light-years; Capella, 43 light-years; Castor, 49 light-years; and Aldebaran, 60 light-years. Being as bright as it is and yet the most distant of these stars, makes Canopus truly one of the most luminous suns in this part of the Galaxy.

February 8 – 14
The Magellanic Clouds

Just 15° and 36° southwest of Canopus, respectively, lie two bar-shaped patches of light that, to the naked eye, look for all appearances like detached pieces of the Milky Way. These clouds are not fragments of our galaxy, but separate, nearby galaxies that astronomers think may be gravitationally bound to our own, like the Moon is to the Earth. They are named the Large and Small Magellanic Clouds, after the Portuguese explorer Ferdinand Magellan, who recorded them in 1519 during his voyage in search of the strait – also subsequently named for him – at the extreme tip of South America.

Because of their proximity to the south celestial pole, the Magellanic Clouds can only be seen from the Southern Hemisphere (though they may be glimpsed very near the southern horizon after sunset from latitude 10°N in late December and early January). As seen from Sydney, Australia, this time of year, the Large Magellanic Cloud (LMC) straddles the southern meridian 55° above the horizon, just after sunset. The Small Magellanic Cloud (SMC) lies about 21° toward the southwest. In a dark sky, they're impossible to miss, and from the Cimmerian seas off the eastern coast of South America, they must have been a startling apparition to Magellan in the early sixteenth century.

Both the LMC and SMC belong to a class of galaxies called 'irregulars,' because they exhibit no definitive shape, unlike those of the spiral and elliptical classes. The LMC is about 33,000 light-years in diameter and lies about 171,000 light-years away. The SMC is about 20,000 light-years across and lies even further away, some 200,000 light-years. Both galaxies are rich in a class of young stars called population I, which are metal-rich stars forged from gas processed in the nuclear fires of previous generations of stars, an older group known as population II stars.

Both galaxies are enveloped in a cloud of cool, electrically neutral hydrogen gas that radio astronomers have traced over 110° of the sky back toward our galaxy. This runnel of gas, called the Magellanic Stream, may have been gravitationally drawn out of one or both of the Magellanic Clouds when they passed too near the Milky Way over 200 million years ago. Some astronomers think the Stream may have been created during the Clouds' first interaction with our galaxy; others think it may be the product of many such interactions.

The Magellanic Clouds are located in an area of sky fairly devoid of bright stars, although these two objects more than make up for this deficit. A pair of 7×50 binoculars is all you need to scan the Clouds for star clusters and nebulae.

In the LMC, the most striking feature is the Tarantula Nebula, designated NGC 2070,* clearly visible to the naked eye at the southeast end of the Cloud. The nebula contains a group of about a dozen very hot, massive stars packed into a region less than a light-year across. This massive stellar bonfire lies at the center of the spider-like tendrils of hot gas that give the nebula its name. If this cluster of hot stars were as far away as Pollux (35 light-years), in our skies it would appear brighter than the full Moon.

In and around the Tarantula Nebula, and, indeed, scattered throughout the LMC itself, are many bright gaseous star forming regions. This area is a hotbed for star birth – and death, as was seen in 1987, when a star exploded on the extreme western end of the Tarantula Nebula. Supernova 1987A was the first supernova seen with the naked eye since 1604, reaching magnitude 2.9 and remaining visible for months. The remnants and outflowing shock wave generated by the supernova are still being studied by astronomers.

Although the demure SMC doesn't exhibit the obvious celestial treasures of its larger counterpart, it nevertheless contains a fair share of star clusters and wispy nebulosity, including the globular cluster NGC 362 and the open cluster NGC 346.

Most impressive is the beautiful globular cluster 47 Tucanae (NGC 104), lying just west of the SMC. Visible to the naked eye as a 4th-magnitude fuzzy star, this object is resolved into a seething ball of stars when seen with a small telescope. Being one of our nearest globular clusters, however, only 20,000 light-years away, makes 47 Tucanae a part of our galaxy, rather than the SMC, as it appears.

* NGC stands for New General Catalog, which is a vast compilation of thousands of star clusters, nebulae, and galaxies.

For those of us in the Northern Hemisphere, it is well worth a trip down under to behold the Large and Small Magellanic Clouds in the same sky with Canopus, Alpha and Beta Centauri, Achernar (Alpha Eridani), and the Southern Cross. Seeing these celestial wonders for the first time is like seeing a different universe.

Also this week:
- Capella, sixth-brightest star in the night sky, is overhead for observers at mid-latitudes in the Northern Hemisphere around 8 o'clock.

February 15–21
A starry hothouse

Where do stars come from? Where did the Sun come from? Planets? Moons? People? The answer can be seen from your backyard this time of year, if you know just where to look.

Go outside around 10 o'clock and face south. There you'll see the great constellation Orion with its two bright signature stars, ruddy Betelgeuse and blue Rigel. Between Betelgeuse and Rigel lies a neat row of three stars running from southeast to northwest – Orion's belt. Now focus on the left or easternmost star in the belt. Its name is Alnitak, from the Arabic *Al Nitak*, the girdle. Let your eye drift southward and slightly westward from Alnitak and you will discern a 'sword' of stars hanging from Orion's belt.

Near the lower end of this sword is a misty patch of light, which can just be seen from a suburban location. The cloud-like quality that you see is not an illusion but the very real effect of looking at glowing interstellar gas. What you're looking at is a fresh stellar bloom, a hothouse where stars are vigorously budding out, their existence just beginning.

This famous object is known as the Orion Nebula. On star maps, it is labeled M42, which just means that it was object number 42 on the list of nebulous objects observed by the French astronomer and comet hunter Charles Messier.*

---

* In the mid-1700s, Messier compiled a catalog of objects that could be mistaken for comets. The Messier Catalog contains a total of 110 objects, although one, M102, is a duplicate listing of M101, and two, M40 and M73, are merely stars.

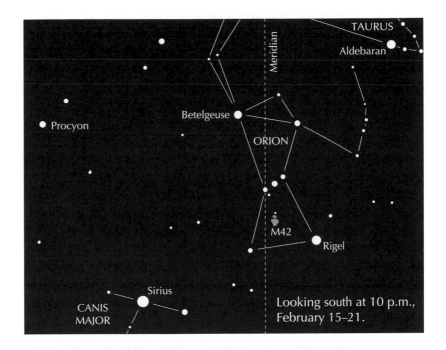

Looking south at 10 p.m., February 15–21.

The Orion Nebula is, without a doubt, one of the finest examples of a bright nebula visible in the Northern Hemisphere, and perhaps in all the sky. Its distance falls somewhere between 1,300 and 1,600 light-years, and yet we can see it with the unaided eye on a dark night. If you have a pair of binoculars handy, by all means aim them at this celestial gem. When you do so, notice how the patch becomes a distinct squarish haze spread over what looks like a knot of glittering stars. Closer inspection with a 3- to 4-inch telescope at low-to-medium magnification reveals a smear of gray light, at the core of which lies a 'fist' of stars. The fist is a gravitationally linked quadruple star system called the Trapezium. (Two other stars associated with the Trapezium can be seen with a larger telescope using higher magnification.) Being among the first stars born in the Orion Nebula, they are the hot, massive, blue suns that are largely responsible for ionizing the hydrogen gas in the cloud, making it glow.

Infrared cameras, which, like night-vision goggles, allow astronomers to 'see' warm objects shining brightly through dense curtains of dust clouds, reveal a family of over 500 newborn stars still swaddled in the warm pockets of dust out of which they condensed. These inchoate stars are packed into a region less than three light-years across, which is equivalent to about two-thirds the distance between the Sun and its nearest stellar neighbor, Proxima Centauri. Images made with the Hubble Space Telescope even reveal well-defined disks of dust orbiting young, emerging stars. These might be incubating solar systems, future places of planets, moons, comets, and, perhaps, life.

By astronomical standards, most of the stars in this gaseous inner sanctum were born only yesterday. A typically cited age is around 500,000 years. The Orion cluster, then, is one of the very youngest and densest clusters known.

Long-exposure photographs capture the compelling nature of the Orion Nebula, the wisps and tendrils that extend across the width and breadth of the entire sword region, but they only hint at its dimensionality. To be sure, the Grand Canyon is beautiful in photographs, but to see it with both eyes from the north or south rim is nothing less than awe inspiring. In three dimensions, the Orion Nebula is, itself, a kind of chasm in the sky. The telescope reveals dark lanes bordering the eastern and northern edges of the nebula. These are 'rims' of dense interstellar dust that actually lie in the foreground. The bright part of the Orion Nebula lies on the 'floor' of this chasm, and there stars have taken root in every crack and crevice, their radiant heat blowing away much of the overlying dust. This allows us not only a privileged view of starbirth, but a semblance of how our origins might have looked over five billion years ago when our own Sun began to flower.

February 22–28 (or 29)
Three star clusters

Auriga the Charioteer (January 25–31) holds more fascination than just ancient star myths. The pentagonal-shaped figure is home to a trio of bright star clusters that, under certain sky conditions, can be seen with the naked eye.

Our galaxy contains hundreds of millions of star clusters, of which we can see mere thousands from Earth. Most star clusters are too faint to be seen without optical aid, but a handful can be glimpsed as misty patches between the stars. Lying in the star-thronged winter Milky Way, Auriga is bedecked with several prominent star clusters, and when the constellation is high in the sky, as it is this week after sunset, its brightest centerpieces, M36, M37, and M38, are easy to locate with just your eyes or binoculars.

These three clusters lie in a row across the southerly half of the constellation. To find them, you must observe in the darkest location you can find, away from the glare of streetlights and buildings. A site in the outer suburbs is okay, but the country is better. The sky must also be free of a bright Moon and high thin cloud cover or haze.

First, locate Capella, the brightest star in the constellation. In the evening sky this time of year it is nearly directly overhead for people in the Northern Hemisphere. Next, go due south from Capella until you reach the star lying in the opposite corner of the pentagon. This is El Nath. (Auriga shares this star with Taurus the Bull, in which it marks the tip of the bull's northern horn. El Nath is from the Arabic *Al Natih*, the butting one.)

Now let your eye move northeast along the second-longest side of the pentagon toward the star Theta (θ) Aurigae. About halfway to this star, you should come across three concentrated patches of misty light running southwest to northeast. Two will lie within the pentagon, but one will be just outside of its southeastern border. This easternmost cluster is M37; the next one toward the west is M36; and the westernmost cluster is M38.

You may find locating these clusters easier with a pair of binoculars. The narrower binocular field of view helps increase the contrast between the clusters and the darker sky background, and with a little magnification you should also notice some differences in the shapes and richness of these clusters. M37 is large and densely packed, with a bright yellow-orange star near its center. It is the 'grainiest' of the three clusters. M36 is much brighter than M37, but smaller and not as rich. M38 is as large as M37, but is fainter and coarser.

The distances to M38 and M37 are about the same, around 4,400 light-years. M36 lies only a little closer, about 4,100 light-years. In the scale of astronomical things, a couple of hundred light-years difference doesn't matter too much. These clusters are neighbors. If M36 were 4,000 light-years nearer, it would appear as bright and as large as the Pleiades.

The clusters differ quite a bit in age. M36 is very young, about 25 million years old, while M37 and M38 are far older, 300 million years and 220 million years, respectively.

While you're in this region of sky, scan with your binoculars northeast and southwest of Auriga. You'll see many more open clusters and star groupings, each containing hundreds, even thousands, of stars. Don't worry about what they are called or how far away or old they are. You can, if you wish, always learn these details later. For now, just enjoy them for their beauty and for giving this region the reputation for being one of the richest areas of the winter Milky Way.

Also this week:

- The trapezoidal constellation Corvus the Crow rises around 7:30 p.m. in the Southern Hemisphere; 11 p.m. in the Northern Hemisphere. In mythology, Corvus was the beautiful white bird of Apollo. The hapless fowl was turned into a crow, however, after he failed to fetch Apollo a beaker of water.

March 1–7
Gemini the Twins

As darkness falls this week, go outside and look high in the eastern sky. You should see two bright stars almost side by side. They are Castor, the westernmost of the pair, and Pollux to its east. Castor and Pollux are the two brightest stars in the constellation Gemini the Twins. If you look carefully, you'll note that Pollux has an orange tint to it, while Castor is white.

Castor and Pollux mark the heads of the Twins. If your skies are clear and fairly dark, you should have no trouble making out a parallel strand of four or five stars trailing southwest of both Castor and Pollux. These stars represent the gangly bodies of the Twins.

Castor and Pollux have been regarded as Twins from antiquity, but the name 'Gemini' dates from the classical period and is a variation of their Italian name, Gemelli. Other cultures had different names for them, but they were always a pair of somethings, whatever they were. The Arabs called them the 'two peacocks,' and the Egyptians the 'two sprouting plants.' The Hindus, evoking a higher plane, referred to them as the 'twin deities.'

In western mythology, the Twins are most prominently linked with the Argonauts, a band of heroes that sailed with Jason in quest of the Golden Fleece. They were supposedly the sons of Jupiter and Leda. Castor was known as an excellent horseman, while Pollux excelled as a soldier and boxer. Roman soldiers believed Castor and Pollux led them to victory in their battles.

According to Martha Evans Martin in her book *The Friendly Stars*, it was common practice in ancient times to swear by the name Gemini. The expression 'Oh Gemini!' is frequently found in Elizabethan literature Evans writes. A corrupted version of this is where we get the expression, 'By Jiminy!'

At the feet of Gemini lies a bright star cluster, M35, that can be glimpsed in binoculars in fairly dark skies. Look just off the last star in the string of stars stretching southwest from Castor, in the direction of the horns of Taurus the Bull. If your eyes are used to the dark and there is no moon or high clouds, you should see a fuzzy patch of light, which is the cluster itself. If you zoom in with a small telescope you will see that M35 is a rich, granular aggregate of 300 or more stars arranged in chaotic strands, like a tangle of Christmas lights bunched up on the floor.

Being the easternmost of the winter constellations, Gemini remains prominent for observers in the southerly realms of the Northern Hemisphere well into the spring months as bluebirds take wing, bass move into shallower water to spawn, and robins hop their way through brush and meadowland. The Twins are also conspicuous in the higher northern latitudes, although winter is still very much in force there. Still, there are signs of the season's change even in those frigid climes: Canadian geese returning from their migration south, creatures arousing from hibernation, and the onset of the unlocking of ice in the rivers. Shadows, once long and dolorous even at noon in January and February, are shorter now, and the Sun seems to rise a little earlier and set a little later each day. And of course Gemini is high in the sky silently trumpeting in spring.

March 8 – 14
The light that hides

Of all astronomical phenomena visible with the naked eye, the zodiacal light is the most sublime. On moonless spring evenings it appears as a soft cone of light no brighter than the Milky Way extending up from the west-southwest horizon well after sunset. Delicate in every respect, dim stars can easily be seen through it, and its most 'brilliant' display is often mistaken for sunset glow, crepuscular rays, or uplighting from a nearby city or town.

The zodiacal light was recorded by Middle Eastern skywatchers in the Middle Ages, but the first known serious observations were made in 1683 by Giovanni Domenico Cassini (the first in the famous Cassini dynasty of astronomers). Cassini, who correctly described the light as a manifestation of a solar cloud of dust, was astonished that something that had been visible since ancient times had been so rarely observed. In 1686, the French science writer Bernard le Bovier de Fontennele echoed Cassini's surprise when, referring to the zodiacal light in his book, *Conversations on the Plurality of Worlds*, he wrote, 'How does a light manage to hide?'

Answer: by being too big, and thus spread out, to be seen. The zodiacal light is nothing more than sunlight scattering off countless particles of microscopic dust orbiting the Sun in a great disk. The disk, however, is so large on the sky that light from the disk is spread thinly, resulting in very low surface brightness. Like the planets, the dust particles are largely confined to the plane of the solar system, or the ecliptic, which is the mean plane of Earth's orbit around the Sun as projected against the stars. The constellations through which the ecliptic passes are known collectively as the zodiac, and it is among these stars that the diaphanous light can be seen.

From our Earthly vantage point, the zodiacal light is broadest near the horizon, but tapers upward at a distance of 60° or 70° (remember, the width of your fist at arm's length is approximately 10° to 12°). Because the glow is aligned along the ecliptic, it reaches a higher altitude and is easier to observe at those seasons when the ecliptic is inclined steeply to the horizon. In the middle to tropical latitudes in the Northern Hemisphere, it is best seen in the west after nightfall in the spring, and in the east before dawn in the autumn. (In the Southern Hemisphere, that would correspond to 'in the west after nightfall in the southern *autumn* and in the east before dawn in the southern *spring*'.)

In the tropics, where the ecliptic is more nearly perpendicular to the horizon, the zodiacal light is visible throughout the year in both the evening and the morning sky. It has also been observed during total solar eclipses immediately around the Sun, blending its milky uniform glow with the glaucous petals of the solar corona.

A manifestation of the zodiacal light also appears in another part of the sky, where it has a different name, the gegenschein, or 'counterglow.' This phenomenon appears as a very faint, elliptical glow some 10° across cast opposite the Sun in what is called the 'antisolar' region. Very dark, clear skies are required to see this elusive feature. One of the more favorable times to view the gegenschein is during the predawn hours in the autumn when it projects against the relatively dull region of the constellations Pisces and Cetus as they set in the west. But dark, moonless spring evenings are also advantageous. Then, it can be seen in the east among the rising stars of Leo and Virgo.

At still rarer times, a band of light can be discerned connecting the zodiacal glow with the gegenschein. This light path extends along the ecliptic and is known as the zodiacal band. Like the zodiacal light, the zodiacal band and the gegenschein both result from sunlight scattering off planetary dust.

To see the zodiacal light, you're going to need dark skies and clear, haze-free conditions. (It is best seen in the country.) A western horizon free of trees and other structures is also advisable. At mid-northerly latitudes (30° N to 35° N), go out an hour and a half after sunset (two hours if you're even further north), let your eyes get used to the dark, and look off toward the west-southwest. Most, if not all, of the sunset glow will have dissipated by this time.

In this 'owl light' look for a soft, white, wedge-shaped glow angled along the constellations of the ecliptic. The wedge will be slightly inclined to the south and may extend up to and beyond the Pleiades. Obviously, because of their limited fields of view, binoculars and telescopes are of no use. You'll have to depend on your eyes and the local weather conditions. Shift your gaze across the sky very slowly, looking for the subtle boundary between the cone of light and darker sky.

Odd as it may sound, you may have better luck seeing it by not looking at it directly. You do this by using the light-sensitive cells in your eyes called 'rods,' which are positioned along the periphery of your retina. This method of observing faint extended objects is called scotopic or averted vision or, more colloquially, 'looking out of the corner of your eye.'

Like sunset colors, halos, and rainbows, the zodiacal light is ephemeral and subtle. Searching for it is a low-key activity that won't inspire everyone. But then again, inspiration is where you find it and what you make of it. Much of astronomy – both recreational and professional – consists of searching for 'light that hides.' Finding that light is what produces the peerless pleasure and personal reward that skywatchers so appreciate, and that too many others, unfortunately, never experience.

Also this week:
- Arcturus rises in the northeast around 8:30 p.m. for observers at northern latitudes. (See May 10 – 16 and May 17 – 23.)

Spring

March 15–21
Spring arrives

If you are looking out of your window this week at piles of dirty snow, a cold rain, or an Arctic front whipping the flags out of the north, you may not believe that spring is at your door, bringing with it the promise of more temperate, benign weather. Just knowing it's spring, though, can be a cheery thought, even during a snowstorm.

Although the spring season lasts approximately thirteen weeks, spring's onset this week around March 21, occurs instantaneously in a quiet, but significant, moment in time called the vernal equinox. The equinox marks the passage of the Sun as it drifts north across the celestial equator – an imaginary projection of Earth's equator into space.

If we could observe Earth from above the Sun's north pole – the direction of which is coincident with Earth's north pole – we would see our little planet plodding along in a counterclockwise direction. Back on Earth, this orbital direction makes the Sun appear to shift a little less than one degree eastward in the sky each day relative to the background stars.* This subtle motion is superimposed on the not-at-all subtle effect of our rotating planet, which makes the Sun appear to move from east to west in a span of several hours.

---

* Contrary to what you may think, the Sun's apparent motion against the background stars is not decisive proof of Earth's annual revolution. A similar eastward drift of the Sun would be seen if Earth resided at the center of the solar system. More conclusive evidence is observed in the slight shift in the positions of the nearest stars relative to the more distant 'fixed' stars at six-month intervals. This phenomenon is called parallax. Still another proof that Earth orbits the Sun is called the aberration of starlight, in which stars appear to be displaced slightly from their true positions in the direction of Earth's motion, just as raindrops appear to fall in toward you at a slant as you walk briskly through them.

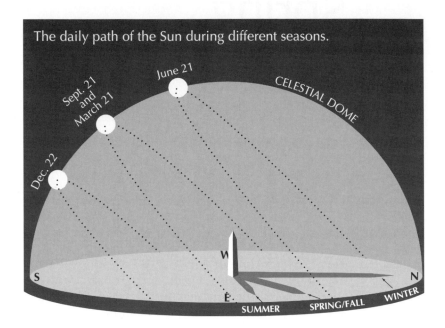

The daily path of the Sun during different seasons.

Naturally, the coming and going of the Sun in such a rapid time frame has more immediate consequences in our lives than its modest east-ward motion. But in astronomy, small particulars have a habit of creating major effects at some point. In school Earth science, we learned that Earth's rotational, or polar, axis is inclined 23.3° from the 'straight-up' position. This inclination, coupled with Earth's annual motion around the Sun, makes the Sun appear high in the sky for half of the year and low in the sky for the other half. The north-to-south motion amounts to about a quarter of a degree per day, except at the Sun's northern and southern limits – the winter and summer solstices – where the motion slows almost to a standstill (in fact, the word 'solstice' means 'Sun still').

Imagine speeding up all this motion so that the Sun crosses the sky once each half second. Each season, then, would last about 45 seconds; a year, 3 minutes. Now we see the Sun racing across the sky with such rapidity that it looks like a continuous arc of light. As the seconds, seasons, and years tick by, it becomes obvious that this arc gradually shifts up and down across the sky like a wobbling hoop. In the Northern Hemisphere, the uppermost reach of the shift is the summer solstice, the lowermost the winter solstice. Poised exactly between these extremes is the celestial equator, where the Sun resides this week.

From all locations in the Northern Hemisphere, the celestial equator arcs across the sky south of the zenith, at an angular distance equivalent to one's latitude. For those in the Southern Hemisphere, it stretches north of the zenith, again by an amount dependent on latitude. So, on March 21, with the Sun centered on the celestial equator, observers standing on the Equator see the Sun overhead at noon, while observers located well north of the Equator, say at latitude 33°N, see the Sun 33° *south* of their zenith at local noon (or 57° above their southern horizon). For those at latitude 40° N, the Sun would be 40° south of their zenith (or at an altitude of 50°) the first day of spring.

The Sun will continue its northward trek until June 21, when it reaches its greatest northern extent 'above' the celestial equator. On that day, the first day of summer in the Northern Hemisphere, the Sun is overhead at noon for people living at latitude 23.3° N, a location known as the Tropic of Cancer, and the northernmost boundary of the Torrid Zone. (Two thousand years ago on the summer solstice, the Sun was located in the constellation Cancer, while the first day of winter found it in Capricornus – hence the names Tropic of Cancer and Tropic of Capricorn.) So if you live at latitude 33°N, the Sun will be 33° 23.3° or about 10° south of your zenith on the first day of summer (at an altitude of 80°). From latitude 40° N it will be about 17° south of the zenith (altitude 73°).

Hence, spring is a good time to determine where the Sun will be in your part of the country at noon on the first day of summer. This is a particularly useful exercise if you're planning to build a deck, install a swimming pool or skylight, or plant a fruit tree. The obvious problem is estimating how much further north 23.3° is in the sky from the Sun's current springtime position at noon. You can determine this simply by using your fist and fingers.

The span of a fist when held at arm's length is roughly 10°. This is true if you have a typical fist. If, on the other hand, you have hefty hands, this measurement may be more like 12°. Calibrate your fist using the Big Dipper, which is well above the northeastern horizon this time of year. The straight part of the handle, from where it is attached to the cup to the bend in the handle (Megrez to Mizar) is very close to 10° in length. If your extended fist blocks out this part of the handle, your fist span is slightly greater than this amount.

After you have correlated your fist size to the Dipper, you only have to subdivide your 10° fist to make up the 3.5°. On a normal fist, this is about two fingers – the index and the medial, again held at arm's length.

Sight in the Sun on the first day of spring around local noon (being careful not to stare directly at it) and, using your well-calibrated fist, offset 23.3°. Note the position with respect to your immediate surroundings. That will be the Sun's maximum altitude for your location at noon on June 20, the first day of summer, when it is highest in the sky and beating down unmercifully. Plan your garden, porch roof, or deck accordingly.

Also this week:
- The Big Dipper is highest in the sky at midnight in the Northern Hemisphere.

March 22–28
Finding true north

You probably know in which direction cardinal north lies from your house, but do you know where celestial north is? True, it's in the same general direction as cardinal north, but celestial north is made with reference to a point in the sky around which all the stars in the Northern Hemisphere appear to pivot. There's an easy way to find celestial north by using the most popular stellar landmark in the sky: the Big Dipper.

The Big Dipper consists of seven bright stars that are a part of the constellation Ursa Major, the Greater Bear. Throughout March, we see this well-known star pattern in the early evening rising cup-first into the sky, with its dogleg-like handle trailing behind. By 10 p.m. this week, the Dipper is high in the northern sky lying cup-down with its handle pointing due east.

By using the two stars at the front of the Dipper's cup as pointers (these happen to be some of the brightest stars in the group) you can find the North Star, Polaris. Simply extend an imaginary line 'downward'

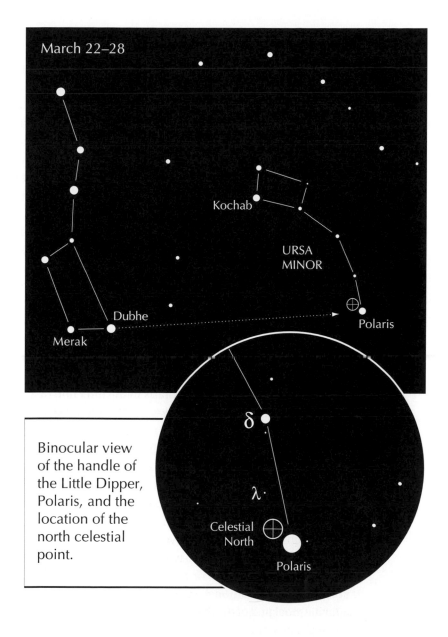

March 22–28

Kochab

URSA MINOR

Dubhe

Merak

Polaris

δ

λ

Celestial North

Polaris

Binocular view of the handle of the Little Dipper, Polaris, and the location of the north celestial point.

from those two stars toward the north horizon. The first bright star you encounter (magnitude 2) from the pointer stars is Polaris.

For centuries, Polaris has been used by mariners and intrepid explorers alike because it is a celestial constant: it never rises or sets. If we could turn off the Sun for a moment and speed up the motion of the sky so that a day was ten seconds long, we'd see all the stars pivoting about this one star in dizzying concentric circles like a rapidly spinning Ferris wheel. This motion, of course, is simply the result of Earth's rotation around its pole. Because Earth's pole happens to point toward Polaris (it has to point *somewhere* in space), the whole sky appears to wheel around that star.

You can dramatically reveal this motion using a typical 35 mm camera and 400-speed film. Simply place the camera on a stable surface or mount it on a tripod, focus on Polaris in the viewfinder, and open the shutter for a half-hour or so. When you develop your film, you'll see colorful concentric star trails centered on Polaris.

If you examine the image of Polaris in your star trail photos, however, you'll notice that it, too, displays a slight arc. Even from the North Star, it's still possible to go further north in the sky.

True celestial north is offset three-quarters of a degree (48 minutes of arc, to be more precise) from Polaris in the direction of Alkaid (Eta Ursa Majoris) – the star marking the end of the handle of the Big Dipper. Although 48 arcminutes may seem like a trivial amount, it looms large in the life of any avid astrophotographer, who must be aligned precisely on this point in the sky in order to take sharply focused, long-exposure photographs of deep-sky objects through the telescope. If the polar axis of the telescope mount is off by even half that amount, the stars in the photographs will appear not as distinct points but as slight lines of trailed light.

The north celestial point will remain in this vicinity of the sky for another thousand years or so, but it is not destined to remain there forever. Earth's rotational axis wobbles slightly owing to the combined gravitational pull of the Sun and Moon on our planet's equatorial bulge. This 'precession,' in turn, causes the true north point to transcribe a great circle in the sky over a period of 25,800 years. Currently, it is drawing nearer to Polaris and will come closest to that star around the year 2100. Fifty-five hundred years from now, Earth's north pole will point in the vicinity of Alderamin, the brightest star in Cepheus the King. Twenty-seven hundred years later, it will point nearest Deneb in Cygnus the Swan, though, even at its closest, that star will lie over 6° away from true north. Thirteen hundred years after that, Delta Cygni will be a little over 2° from true north. And 2,000 years after that – or in 13,500 – true north will incline toward Vega in Lyra the Lyre, though it will be offset by about 5°.

During the age of the buidling of the Egyptian pyramids, 4,600 years ago, Thuban, the brightest star in Draco the Dragon, occupied the title of North Star. It will do so again around the year 23,300.

Also this week:
- Sirius is on the meridian after sunset, as seen in the Southern Hemisphere.

March 29–April 4
Finding true south

The Northern Hemisphere has the bright North Star as its guiding light to astronomers, navigators, surveyors, and Boy Scouts everywhere. Not so the Southern Hemisphere. The place in the sky toward which Earth's south polar axis points appears virtually starless.

Except, that's not exactly true. Sigma (σ) Octantis, which can just be seen with the unaided eye in a dark sky, is designated Polaris Australe, the South Pole star. Sigma, which resides in the dim constellation Octans the Octant, presently lies 1.03° from the pole, but unlike the north celestial pole star, it's not getting any closer. By 2010, this distance will have increased by about two arcminutes. Sigma Octantis was nearest the south celestial pole around 1830, when it was only 45 arcminutes away.

There are, however, nearly half a dozen other stars that lie still closer to true celestial south. Problem is, they're too faint to be seen with the naked eye. The brightest is about magnitude 7, the others fainter still.

So how do you find true south without a bright stellar beacon to guide you? This time of year, Crux, the Southern Cross, is high in the evening sky in the southeast. By using the longest axis of the cross – marked by

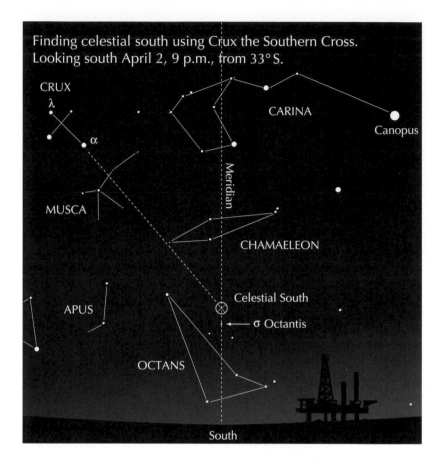

Finding celestial south using Crux the Southern Cross. Looking south April 2, 9 p.m., from 33° S.

the stars Gamma and Alpha Crucis – as both pointer and measuring stick, you can hop your way to the south celestial pole.

The apparent angular distance between Gamma and Alpha Crucis is 6°. Polaris Australe lies about 27° away, or four and a half times the length of the Cross. This amounts to about three fist widths held at arm's length. A quick check of this region in binoculars should reveal a star field similar to the one illustrated here, which you can use to pinpoint the south celestial pole. (You will likely have to tilt the chart to orient yourself to the stars' positions as they appear in your field of view.)

When the Southern Cross is not above the horizon, say around early November, the task becomes a bit more challenging because there is no prominent star pattern to use as a guide. In this case, you must look for the 2nd-magnitude star Ankaa in Phoenix and the slightly fainter Beta (β) Hydri in Hydrus. These two are your pointer stars. A line extended from Ankaa through Beta Hydri, and thence onward another 12° (slightly over one extended fist-width), puts you very near the pole.

Just as precession shifts Earth's axis away from pointing at the North Star in the Northern Hemisphere, so too will it induce the south polar axis to incline away from Polaris Australe. By the year 5700, 3.3-magnitude Omega (ω) Carinae will be the new South Star. By 7700, Iota (ι) Carinae assumes that position, followed by the brightest south pole star in the entire precessional cycle, Delta (δ) Velorum, in the year 9000.

Also this week:
- The blue-white star Spica in Virgo rises after sunset as seen in the northern sky.

April 5–11
The Beehive Cluster

In the southern part of the Northern Hemisphere this week, spring begins to manifest itself in fields and meadows as wild flowers bloom and tender new buds push out into the sunlight. You may see hummingbirds and bees working the blossoms of morning glories and trumpet vines. How appropriate, then, that near the apex of the night sky this week is a star cluster just visible to the naked eye that goes by the name, the Beehive Cluster.

Located in the faint constellation Cancer the Crab, the Beehive, also known by its more ancient name, the Praesepe (the Latin root for manger), is one of the nearest and largest open clusters. It can easily be seen with the unaided eye in a dark sky as an oval glow over 1° across (the width of two full Moons side by side). Closer examination with binoculars or a low-power telescope reveals an aggregate of stars, like a chaotic throng of buzzing bees.

Being bright enough to observe without the aid of a telescope, the Praesepe is one of the few star clusters mentioned in ancient literature. Richard Hinkley Allen, in *Star Names: Their Lore and Meaning*, writes that the Greek astronomer Hipparchus, around 120 B.C., described it as a

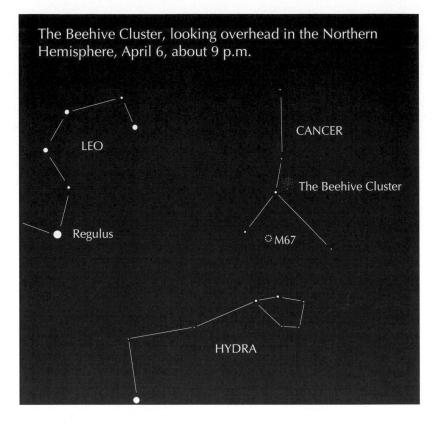

The Beehive Cluster, looking overhead in the Northern Hemisphere, April 6, about 9 p.m.

LEO

CANCER

The Beehive Cluster

Regulus

M67

HYDRA

'Little Cloud.' An older reference comes from Aratus of Soli (court poet of Antigonus I, king of Macedonia) who, around 250 B.C, referred to it by a similar name, the 'Little Mist.' Throughout ancient times, from Homer to the Roman scholar Pliny, the Praesepe was allegedly used to indicate changes in the weather. If the cluster appeared bright and sparkling, the weather would be fair. If, however, it appeared darkened in a clear sky, this was taken to mean that a storm was in the offing.

Of course, no one then knew that the Praesepe was in fact a star cluster until Galileo turned his telescope on the object in 1610, revealing a tangle of 36 bright stars. In fact, modern counts bring the number up to nearly 400, though most of these stars are extremely faint. Any small telescope at low magnification should reveal between 100 and 150 members. Because of the cluster's large apparent diameter, binoculars may present a more pleasing view than a telescope.

While you're in the area, be sure to look 9° south of the Beehive – or about 2° west of Alpha (α) Cancri – for another star cluster: M67. This is a rich, compact cluster spanning about ½° of sky and containing several hundred stars fainter than magnitude 10. It's visible in binoculars as a distinctly granular patch of gray light. In any modest sized telescope, you should have no trouble seeing at least 60 of the brighter stars. M67 is one of the oldest known open clusters in the Galaxy: about 4 billion years old, which is approximately the same age as our Sun.

Also this week:
- The Southern Cross is on the meridian around midnight for southern observers.
- Bright meteors seen this week and next may be attributable to the minor meteor shower the Virginids. This shower exhibits several maxima during its active period throughout March and April, during which from two to five meteors per hour may be seen. Virgo, from which the meteors appear to emanate, rises fully by 10 p.m., though more meteors will likely be seen after midnight.

April 12–18
A kingly star

An old starwatcher's saying goes, 'Poke a hole in the bottom of the dipper's cup and the milk will spill on the lion's back.' Indeed, if you follow an imaginary stream of milk southward from the bottom of the bowl of the Big Dipper, you'll dribble right onto the back of Leo the Lion.

You can't miss Leo this week, especially with his distinctive mane that resembles a sickle (others describe it as a 'backwards question mark'). At the end of the sickle's handle shines a lone jewel, Regulus, a star whose name is a Latin derivative meaning 'the prince,' or as others interpret it, 'the little king.'

Though it is the eighteenth-brightest star in the sky Regulus looks a bit lost, surrounded as it is by darkness and stars far fainter than itself. It ranks in brightness a notch below white Deneb in Cygnus, seen in summer skies, but is brighter than orange-hued Pollux in Gemini, which, even though it is Regulus's nearest bright neighbor in the sky, lies over 30° to the northwest. On a dark night with the star highest in the sky, Regulus looks predominantly white in color, but is tinged with just a hint of blue. Near the horizon, it may appear reddish, because its light bends through a greater thickness of Earth's atmosphere at that altitude.

If you note the comings and goings of Regulus, you'll find that this spring star doesn't belong to any one season. It belongs to them all. Regulus first appears in the sky New Year's evening, winking at us through bare trees around 10 o'clock. By late February, it's in the sky all night, rising when the Sun sets and setting when the Sun rises. Regulus remains visible until mid July, when it sets soon after the Sun. Thereafter, it vanishes in the Sun's glare in the west, but reappears in time for fall's arrival in late September, when it peeks over the eastern horizon just before dawn.

When you get right down to it, there are a number of these 'season-less' stars and even seasonless constellations. For example, another spring star, Arcturus – fourth-brightest star in the sky – appears in the east the first of March at sunset and doesn't vanish from the evening skies until late November. And the summer constellations Cygnus, Lyra, and Aquila hang around in the western sky until well after the first day of winter.

Traditionally, however, bright stars and constellations are associated with the seasons in which they are highest in the sky. Until Arcturus takes the prince's place, followed by the bright stars of summer, Regulus remains the crowning star of spring.

April 19–25
The surprising Lyrid meteors

This week, Earth encounters the richest part of the Lyrid meteor stream, providing us with the opportunity to see a spritzing of 'shooting stars.' The best time to observe this shower is in the after-midnight hours, when the night side of Earth faces into its orbit, thus sweeping up more meteoritic debris. Avoid streetlights and brightly lit billboards, as their glare masks the fainter meteors. The Moon, too, can overwhelm the brief flash of a meteor, though if it displays no more than a crescent phase and is low in the sky, it should not pose too much of a problem.

Meteor showers are not really 'showers' at all, but an increase in the number of meteors seen per hour coming from a particular point in the sky, called the radiant. In the case of the Lyrid meteors, their paths, if traced backward on the sky, point in the general region of the constellation Lyra – hence the name 'Lyrids.' Meteors seen at this time coming from other directions are not associated with the Lyrids and are referred to as 'sporadics.'

Meteoroids – the stuff that actually causes the brief streaks of light in the sky during meteor showers – essentially consist of comet debris. When a comet ventures into the inner solar system, the Sun's warmth boils off the volatile ices and rocky particles that comprise the comet's crust, leaving a trail of comet crumbs. As Earth passes through the comet's detritus, some of the tiny bits of dirty ice burn up in the atmosphere, where they may be seen from the ground as shooting stars.

This week's meteor shower appears to radiate from a region of the sky between the constellations Lyra and Hercules. Lyra, a parallelogram of four stars with the brilliant blue-white star Vega near one corner, is overhead around 4 a.m. The radiant itself is located west-southwest of Lyra. In most years, observers can expect to see less than ten meteors per hour. Take note, however, that the rate can be significantly higher. A great display was seen in 1803 and once again in 1922. The shower plateaued in activity for the next 59 years, but picked up again in 1981. And in 1982, American observers reported between 75 and 90 meteors per hour around the peak. The shower's unpredictability makes the Lyrids an event not to miss.

April 26–May 2
The Big Dipper's many identities

In England it is known as the Plow (or 'Plough,' to match the proper English spelling). In Southern France, it has been called the Casserole or Saucepan. Germanic people refer to it as the Great Wagon, while the Chinese call it the Northern Ladle. In the United States, skywatchers and nonskywatchers alike know it as the Big Dipper.

Though it's not officially one of the 88 constellations (it is an asterism, or star grouping, marking the flank and tail of Ursa Major, the Greater

Bear) it is, nevertheless, one of the most asked-after and recognizable star patterns by both children and adults. It is often one of the first sky objects to attract beginners to astronomy, probably because it is circumpolar in the Northern Hemisphere for latitudes north of the 41st parallel, and thus seen prominently throughout the year. You may have spied the Big Dipper yourself high in the northeastern sky this week around 9 o'clock.

Ursa Major is one of our oldest constellations. In ancient times, it was comprised of just the seven stars that make up the Dipper: Alpha (α), Beta (β), Gamma (γ), Delta (δ), Epsilon (ε), Zeta (ζ), and Eta (η) – but was later enlarged to include a dozen other bright stars to make up the main bear figure. Two thousand years ago, the Dipper looked virtually the way it does today. In fact, you'd have to go back nearly 90,000 years to see significant changes in its shape. So the familiar pattern of stars we know so well has been around since the dawn of civilization. Its prominence in the sky has made the Big Dipper a kind of free-association test for different civilizations down through the ages.

Many people are surprised to hear that the Greeks were not the only civilization to associate a bear with the Dipper (a long-tailed bear at that). The bear theme can be found in the mythologies of the Arabians, the Hebrews, and various North American Indian tribes, particularly the Cherokee and Iroquois Indians. But not all cultures have ascribed a shaggy omnivore to the Dipper. In *The New Patterns in the Sky,* Julius Staal writes that the Babylonians saw a supply wagon, while in north-west Europe, people fancied a plow. The Romans envisaged seven plowing oxen, while the Egyptians fashioned a bull's hind leg from the cup and handle (a 'thigh in the sky,' as it were).

The Dipper predictably took on the shape of the beasts and artifacts familiar to a particular region. In North Africa, people saw a camel, while those in the East Indies depicted the pattern variously as a shark or a canoe. The Sioux of central North America saw a pesky skunk in the Dipper's pattern, but the Maya Indians of Mexico projected an even more unsavory image – the evil god, Hunracan, who was able to go any-where at will causing war and chaos. It is from this name that we derive the word 'hurricane.'

In Northern India, the seven stars were collectively called Saptarshi, or the Seven Sages. Moreover, the Hindi recognized that one of the seven, the star in the bend of the handle, which they named Vasishta, had a wife. She can be seen to this day, though small and faint, standing next to her beloved. According to Hindi mythology, Vasishta was the teacher of Rama, who was worshipped as an incarnation of Vishnu in Hinduism. (We know the brighter star as Mizar and its fainter companion as Alcor.)

Apparently, the Big Dipper has been many things to many people: bears, plows, saucepans, sages, and gods. But the Dipper is also something else to astronomers. With the exception of Alpha and Eta – the first star in the cup and the last star in the handle – the inner five stars of the Dipper share the same motion through space. In other words, they are a star cluster. Astronomers call it the Ursa Major cluster. It includes some 100 stars (including Sirius) spread out over the sky, each moving in approximately the same direction in space.

So you can also say that the Dipper makes up the heart of the nearest star cluster to Earth, the center of which lies about 75 light-years away. Their collective motion, however, means that in 100,000 years, the familiar dipper shape will be largely unrecognizable. As the inner five stars gradually move together like a school of fish, Alpha will be left behind, lying even further east of the main group than it does now. Eta, too, will slip more under the handle. The new pattern will look more like a pointy high-heeled shoe. One wonders what civilization (in whatever form it may exist 100,000 years from now) will call the Big Dipper then?

Also this week:
• Alpha and Beta Centauri straddle the meridian at midnight in the southern sky.

May 3–9
The Eta Aquarid meteor shower

Practically on the heels of the Lyrid meteor shower (April 19–25) follows one of the best meteor showers of the spring: the Eta Aquarids. Like the Perseids in August and the Geminids in December, the Eta Aquarid shower gets its name from the area of the sky from which the meteors appear to radiate – in this case, a star designated by the Greek letter Eta (η) in the constellation Aquarius the Water Bearer.

The Eta Aquarids first appear around April 21, and some can be seen until May 21. But the shower's peak occurs around May 4. Though Aquarius rises about 2:30 a.m. this week, meteors may be seen higher in the sky as well as to the north and south of the constellation's location at that time. You may want to wait until 3 or 4 a.m. when the radiant is well above the horizon.

The average number of Eta Aquarid meteors seen per hour is listed as 60, but you're not likely to see that many, especially if you're located in a brightly lit suburban area or if there is a beaming Moon in the sky. Under such conditions, you may see only a handful per hour. If you scan the sky with binoculars, however, you might see a few zip by in the field of view. Also helpful is the fact that many of these meteors are bright yellow, and thus easier to see from moderately light-polluted skies. A few will likely leave brief smoke trails in their wake.

From a dark-sky site, you may be able to see between 20 and 30 meteors per hour, particularly if the shower's peak time occurs when it is dark on your side of the planet.

Also this week:
- Vega rises an hour or so after the Sun this week for those at northern latitudes.
- In the Southern Hemisphere, the Southern Cross is on the meridian about 9 o'clock. (Observers in the Northern Hemisphere see May 24 – 30.)

May 10–16
Boötes the Herdsman

High overhead as darkness falls in the Northern Hemisphere these early summer evenings shines a lone, bright orange-hued star. Just go out and look almost straight up around 9 o'clock; you can't miss it. This star is Arcturus, the luminary of the constellation Boötes the Herdsman.

If you live in a suburban location, you can probably just make out the constellation's star pattern. It looks like an ice-cream cone with bright Arcturus at the cone's apex and the ice-cream end pointing more or less toward the northeast. Martha Evans Martin, in her book *The Friendly Stars*, was not much impressed with Boötes, however. 'It is a very pretty constellation,' she writes, 'and of some astronomical interest, but it is not remarkable to the ordinary view except for its one brilliant star.'

What it may lack in aesthetic beauty, however, it more than makes up for in heritage. Boötes, like Ursa Major, is another of one of our oldest constellations, and as a result, its mythology is considerably rich and varied. The name is a derivation of the Sumerian *Riv-but-sane*, the Man Who Drives the Great Cart. It is associated with the farmer who plows his fields in the spring (when Boötes first appears). The Romans referred to Boötes as the Herdsman of the Septemtriones, the seven oxen, which are represented by the seven stars of the Great Cart, known today as the Big Dipper.

According to one Greek legend, Boötes is actually Arcas, the son of Zeus and the nymph Callisto. Arcas suffered many hardships after being robbed of all his possessions. In order to make his living, he invented a plow that was drawn by two oxen. Like most mothers, Callisto was so proud of her son's resourcefulness she wanted to tell the whole world. She convinced Zeus to place her son in the sky together with his plow.

Giuseppe Maria Sesti, author of *The Glorious Constellations*, asserts that Boötes' mythological tree can be traced back 8,000 years, to a time when the constellations were in different positions in the sky. Over a period of several thousand years, precession, the wobble in Earth's rotational axis, noticeably changes the positions of the stars with respect to the north celestial pole. Since all the stars in the Northern Hemisphere appear to revolve around the north celestial pole, a star located near the pole is of considerable importance (as Polaris is today). It also stands to reason that stars near the pole would also be assigned some important role.

If we look back in time to where the north celestial pole was nearly 7,000 years ago, around 5000 B.C., we find it some 12° north of the head of Boötes. Moreover, with the celestial sphere skewed around so, the constellation would have been seen by the people of that time as standing upright on the northern horizon at midnight on the summer solstice, apparently holding up the vault of the heavens near the point of the (then) north celestial pole. Sesti believes that the only mythological figure worthy of this important position was Atlas the Titan. Boötes' right arm, arched above his head in support of the sky, may have consisted of the semicircle of stars known today as Corona Borealis.

The gradual movement of the north celestial pole among the stars has since shifted the spotlight away from Boötes, forcing the constellation to assume a more agrarian position in the sky. We'll have to wait nearly 19,000 years for Boötes to once again assume its classical pose as Atlas.

Skywatchers who have access to a telescope should look for M3, a rich, condensed globular cluster located about 12° northwest of Arcturus in the adjacent constellation Canes Venataci, the Hunting Dogs. It is one of the most beautiful globulars in the northern sky, next to M13 (see August 2 – 8). In binoculars it can just be seen as a fuzzy star.

May 17 – 23
A star that bears watching

By midnight, Boötes hangs from the top of the sky like some immense starry mobile. What so attracts the eye in this cone-shaped constellation, however, is the brilliant orange star Arcturus, fourth-brightest star in all the sky. Its name is Greek for 'bear watcher.'

Proximity is one reason Arcturus is so bright. The star is only 34 light-years away, which, in terrestrial terms, is a little over 200 trillion miles. Daunting as this distance may sound, it is a stone's throw across an astronomical pond. Another reason Arcturus is bright is because of its size, which relates directly to how much surface area is available to radiate light. Though its temperature is 4,200 kelvins,* or about 1,600 kelvins cooler than our Sun, its diameter is 25 times that of the Sun, which makes it nearly 22 million miles across.

Such a star, if placed where our Sun is, would be a shimmering red sphere with an apparent diameter of 12.5° in our sky – the width of 25 full Moons placed edge to edge. Earth would not be a pleasant place to live: Death Valley, where temperatures typically soar over 120 degrees Fahrenheit (50 °C), would be an oasis compared to what Earth would be like near such a star.

* The Kelvin scale corresponds to the Celsius scale but assigns zero degrees to absolute zero, or −273 degrees Celsius. In the Kelvin scale, room temperature (72 degrees Fahrenheit) is about 300 kelvins. The Sun's surface temperature, 5,800 kelvins, is equivalent to about 10,000 degrees Fahrenheit.

One of Arcturus' unique properties is its motion through space. The measure of a star's apparent motion in the sky over a year's time (in other words, how much it moves with respect to the more distant, and 'fixed,' background stars) is called the annual proper motion. The annual proper motion of a star includes both its angular rate of motion (in arcseconds per year) and its direction of motion in the sky.

When astronomers measure Arcturus' annual proper motion, they note the star shifts its position on the sky by 2.3 arcseconds per year in the direction of the constellation Virgo. Although this may seem like a very slight amount – the full Moon covers ½°, or over 1,800 arcseconds of sky – it's a remarkable amount for a star. In 5,000 years, Arcturus will have moved 3° on the sky in the direction of the constellation Virgo, equivalent to six apparent lunar diameters. This makes Arcturus one of the fastest moving first-magnitude stars in the sky (only Alpha Centauri moves faster).

The stars may look frozen in time, but they are not. If we could reduce several million years of proper motion to a minute's time, we'd notice that *all* stars move. There is no such thing as a 'fixed' star. Like Van Gogh's *Starry Night*, the heavens would be streaked with arcs of stars racing across the sky in all directions and at different speeds. Arcturus, though, would be one of the fastest moving stars, dashing across the heavens to become, in about 20,000 years, a new member of the constellation Virgo.

Also this week:
- Observers in the Southern Hemisphere may see the three stars in the end of the handle of the Big Dipper just above their northern meridian this week around 9 o'clock.
- Vega in the constellation Lyra the Lyre rises just before sunset at northerly latitudes and can be seen in the northeast at dusk.

May 24–30
Quest for the Southern Cross

When I first learned the constellations as a boy growing up in Corpus Christi, Texas, (latitude 28° N), I was intrigued by the fact that there was a sizeable portion of the sky below my southern horizon that I never saw.

The Southern Cross, looking south from latitude 25° N on May 24–25.

Rumor had it, though, that from my latitude you could see the northern-most stars of the Southern Cross, also known as Crux, in mid-April around 10 o'clock.

I confirmed this for myself one clear April evening from the flat grain fields west of the city. In case you don't know, portions of coastal south Texas are remarkably flat and, save for the occasional red-blinking antenna beacon and lonely mercury-vapor yard light, horizons in the country are generally unobstructed. It is not unusual, when the air is transparent, to see stars within a degree or two of the horizon line itself.

In fact, there are many accessible locations in the Western Hemisphere where one can see a few extreme southern stars: south Texas, the Florida Keys, the Baja Peninsula, the west coast of Mexico, the Yucatan, and locales in the Caribbean and Hawaii.

If you find yourself in one of these southerly regions this spring, you might begin your southern sky quest by looking for the northernmost stars of the Southern Cross. The torqued square of stars forming Corvus the Crow, which hangs fairly high in the sky due south around 10 o'clock this week, makes an excellent signpost for this southern group. The Cross lies 40° below (or due south of) Corvus. Thus when the Crow flies high, the Cross is nigh. Just lower your sight to just above the horizon and look for a triangle of stars – the three northernmost stars of the Cross. (Be sure you have an unobstructed horizon.) The brightest star of the three, lying furthest east, is Mimosa, named after the tropical shrub. To see the Cross's brightest and southernmost member, Acrux, you will need to be south of latitude 25° N.

If you wait another hour and a half, you should be able to see the bright pair Alpha (α) and Beta (β) Centauri climb to take the Southern Cross's place. (If you're at latitude 28° N, however, you will only be able to see the northernmost-placed Beta, also known as Hadar, as it peeks briefly above the horizon.)

Alpha and Beta Centauri are ensconced deep in the southern Milky Way. Scan this area and the region north and northwest with binoculars, and you will see many 'clouds' of stars. One very noticeable object looks like a star to the naked eye, but in binoculars appears distinctly 'woolly.' Closer inspection with a telescope reveals this to be the great globular cluster Omega (ω) Centauri, a sphere of millions of gravitationally bound stars, and one of the most magnificent sights in the sky.

These are only a few jewels spilled from a chest of riches. The southern sky harbors countless bright stars, clusters, nebulae, and, most notably, the Large and Small Magellanic Clouds, the Milky Way's neighboring galaxies (February 8 – 14). If you can't travel to the Southern Hemisphere anytime soon, the next best thing is to find yourself in the southerly realms of the Northern Hemisphere this spring where you can catch tantalizing sights of the celestial vistas on the 'other side' of the universe.

Also this week:
- Around 10 o'clock, look south of the Big Dipper's handle for a lone, tenuous 'cloud' some ten full-Moon widths in apparent diameter. Binoculars reveal a few parallel threads of stars all about the same

brightness. This is the Coma star cluster, one of the best known and well-studied open star clusters in the sky, because it is so nearby (about 250 light-years). The star strands make up the 'hair' in the constellation Coma Berenices, Berenice's Hair.

- The Scorpiid/Sagittariid meteor shower is active between April 15 and July 25 but peaks during this week each year. The shower can yield as many as 20 meteors per hour, although only under optimal viewing conditions. It is also known to produce the occasional fireball. The radiant is highest around midnight.

May 31–June 6
A cosmic window

Practically overhead this week at 9:30 p.m. is a dark, dim region of sky that figures very prominently in astronomical observation and research. To the naked eye its appearance is undistinguished. There are no bright stars or distinctive constellations here, but it is bracketed by four prominent star groups, which make it easy to isolate.

To its west we find the constellation Leo the Lion, just beginning his descent toward the horizon. You can't miss the sickle-shaped asterism that forms the lion's head, with Regulus as the handle star. To the east is ruddy Arcturus in Boötes, brightest star in the northern nighttime sky. High in the north, the Big Dipper lies cup down, its crooked handle pointing back toward Arcturus. And if we continue on that arched course through Arcturus further south, we come to yet another star, blue-white Spica in Virgo the Virgin, which is approaching the meridian.

Practically on the meridian at this time is a point in the sky called the North Galactic Pole, a point marking the north pole of our galaxy. (There is also a South Galactic Pole, which is in the opposite part of the sky in the constellation Sculptor.) The North Galactic Pole lies a little north of a line between Arcturus and Denebola, the star representing Leo the Lion's tail. You won't find a bright star marking the galactic pole's location, although, if your skies are dark enough, you will see a faint scattering of stars in this proximity, marking the rarefied constellation Coma Berenices.

This unassuming piece of sky is our cosmic window onto a universe

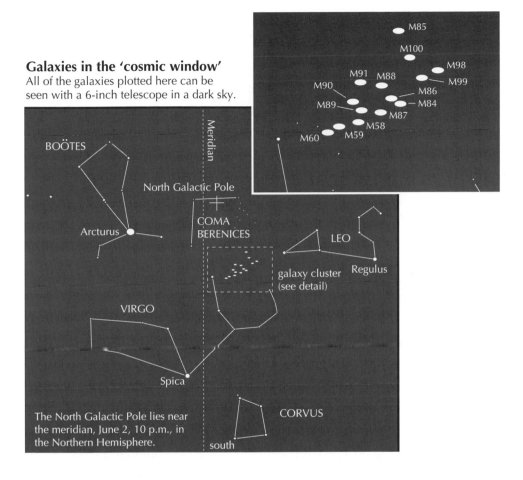

Galaxies in the 'cosmic window'
All of the galaxies plotted here can be
seen with a 6-inch telescope in a dark sky.

that lies well beyond the confines of our galaxy. The reason this area
appears so starless is because our line of sight passes through the thin-
nest part of the Galaxy, in which there are very few stars between us and
the more remote universe beyond.

As mentioned earlier (January 11 – 17), our galaxy is shaped something like a wafer. When we look through the plane of the wafer, as we do
during the summer and winter months, we see lots of stars. The luminous band of the Milky Way stretching across the sky is essentially our
view through the nearby portions of this disk. In the spring and fall
months, however, our nighttime view is perpendicular to the plane of
the Galaxy. It's like looking up at the sky from a clearing in the forest

rather than looking through the dense trees. When we look in this direction, our view is out of the thin part of the disk and deep into intergalactic space.

Although it may look as if there's nothing of interest to see in this direction, close scrutiny with a 6- to 10-inch telescope at low magnification reveals otherwise. On a clear moonless night away from the light glow of the city, slowly scan the region east and a little south of Leo's hindquarters and you will occasionally see faint, fuzzy patches of light interspersed among the stars. Sometimes, you may even see two and three together in the same field of view. These ghostly glows – some round and some oval – are actually other galaxies, tens of millions of light-years away. Each contains hundreds of billions of stars. It is conceivable that some of these stars may have planets and that a select few planets may harbor lifeforms of one type or another. And there are billions of galaxies in the universe.

Because the light we see from these galaxies tonight actually left them tens of millions of years ago, this view forms a kind of cosmic window in time as well as space. The distance the light traversed is so great that it is only now reaching us. We see the galaxies as we might see dinosaurs walking across ancient continents in a time before they vanished from the face of the earth.

June 7 – 13
A summer place

Looking toward the east this week after dark, we see a few bright stars making their way into the sky. These belong to the first string of summer constellations that in another month will assume their place overhead, where the spring stars are now.

We've already mentioned Arcturus in Boötes (see May 10 – 16 and 17 – 23), which is practically at the zenith by nightfall. I like to think of Boötes and its adjacent constellation, the little half-circlet of stars to the north known as Corona Borealis, as bridging the spring and summer seasons, because they are prominent in both.

Coming into view around nightfall is Vega of the constellation Lyra the Lyre. Vega is the brilliant blue-white star rising well above the tree-

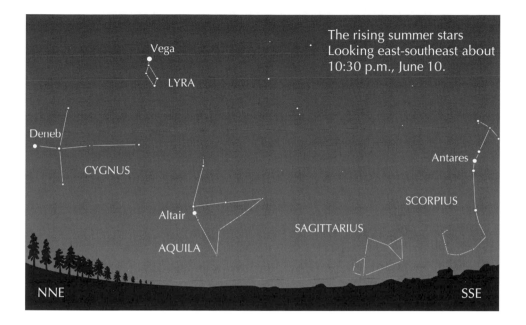

tops toward the northeast. Two and a half times the size of our Sun, Vega is also twice as hot, about 10,000 kelvins. Vega stands near one corner of a group of stars that form a faint parallelogram, which represents the lyre. On July 16, 1850, Vega became the first star ever to be photographed through a telescope. The photograph was made using the 15-inch refractor at the Harvard Observatory in Massachusetts.

More recently, images made at radio wavelengths of Vega show it to be enshrouded in faintly glowing dust, evidence, some astronomers assert, of a budding planetary system. However, given the short lifespan of hot stars like Vega (about 300 million years) planets will likely not have enough time to form. Compare this with our solar system, which took over 4 billion years to condense into planets.

Northeast of Vega is Deneb, brightest star in Cygnus the Swan (also called the Northern Cross). Deneb is one of the largest known super-giant stars. It is about 60,000 times more luminous than the Sun and 25 times as massive. No wonder Deneb is so bright, even at the remote distance of 1,500 light-years.

This Cygnus region is a fascinating one to explore in binoculars since it lies in the thick of the summer Milky Way. By this time next month, it will be higher in the sky, free of the dimming effects of the atmosphere near the horizon, and well placed for viewing. (See July 12 – 18.)

A little after 9:30 p.m., an angry red star rears up far to the southeast. This is Antares in Scorpius. Antares, Greek for 'rival of Mars,' lies at the heart of the mythical scorpion and was one of the royal stars of ancient Persia. In China it was known as the 'fire star.' This reddish star is classified a red giant by astronomers. If Antares were placed where the Sun is now, its huge size would easily swallow up Earth.

Another bright summer star finally breaches the horizon around 10:30 and comes fully into view by 11 p.m. Altair is the brightest star in the constellation Aquila the Eagle. With a distance of only 16 light-years, it's the nearest of this group of stars. Altair is one and a half times larger than the Sun and ten times as luminous. Intriguingly, Altair rotates on its axis once every six and a half hours (compare this with the Sun, which rotates once every 25 days). Because of this rapid rotation Altair, as seen from a hypothetical planet in orbit about the star, would look like a flattened sphere.

If your sky is free of city light pollution, you should also be able to make out the Milky Way, paralleling the horizon northeast to southeast forming a kind of dusty 'backdrop' to the incredible summer stars.

Summer

Summer officially arrives in the Northern Hemisphere this week, in the form of the summer solstice, the instant the Sun reaches its most northerly position for those living north of the equator. In the Southern Hemisphere, meanwhile, the Sun reaches its lowest point in their sky, hence this week marks the beginning of southern winter.

On the first day of summer, the Sun is directly overhead at noon for people living at latitude 23.3° N, a region denoted as the Tropic of Cancer. There's nothing cryptic about this value. It's just how much Earth's rotational axis is tilted with respect to the plane of the solar system. If Earth's axis were tilted more, say 30°, the Sun would be overhead at that latitude instead.

As we orbit the Sun, Earth's tilted axis makes the Sun appear to gradually move north to south and back again from any given location in the world. But only in regions within 23.3° of the equator does the Sun actually ever stand directly overhead. If you live at latitude 30° N, the Sun is still 6.7° south of your zenith on the first day of summer (30°−23.3°); from latitude 40° N, the Sun is 16.7° south of the zenith on that day. (For more on Sun angles, see March 15–21.)

Since the vernal or spring equinox last March, the Sun has crept ever northward, averaging about one degree every four days, or the apparent width of two Suns side by side. This motion gradually changes the Sun's angle of light with the ground, as well as where sunlight falls in your region of the world. In the Northern Hemisphere, the Sun is very high in the sky at noon on this date; noontime shadows are short. The opposite conditions prevail in the Southern Hemisphere at this time of year.

Perhaps you've noticed sunlight streaming through a window in winter that remains shaded in summer, or that the Sun rises and sets further north of prominent horizon landmarks than it did in March.

The Sun does not move at a constant rate in its north-to-south oscillation. This is particularly obvious by noting its sunrise or sunset positions on the horizon over a period of days around the equinox and the solstice. The Sun moves fastest around the equinox, almost half a degree a day at mid-latitudes. But like a pendulum at the limit of its swing, it slows almost to a complete stop around the time of the solstice, rising at the same place on the horizon for almost a week. The word solstice, in fact, comes from the Latin *solstitium*, meaning, 'Sun stand still.'

The Sun's motion was closely followed by a number of North American Indian tribes. The Pueblos feared the Sun might cease its back-and-forth swing at either extreme – the summer or winter solstice – creating disastrous consequences for their climate and agriculture. The Hopi relied on the Sun's horizon position to tell them when to plant crops, when to harvest, when to hunt, and conduct weddings and other ceremonies. They kept track of the Sun, and thus the season, by noting its position relative to sharp peaks and mesas on the horizon, in effect, creating a horizon calendar.

The summer solstice has been called the 'noon of the year' because the Sun is at its highest overhead postion for the year (although this reflects a Northern Hemisphere bias; for the Southern Hemisphere, it must be the 'midnight' of the year). In the Northern Hemisphere, the summer Sun is higher in the sky with respect to our horizon, and therefore remains above the horizon for a longer daily period than it does during the winter months.

When we think of the first day of summer, our thoughts likely turn to picnics, trips to the lake or beach, baseball, and general lazing about. Now's the time to get out and enjoy the weather before the first chills of autumn return, which always seems just around the corner for those living in the more northerly latitudes of the world.

Also this week:
- The upper portion of Scorpius, including the bright red star Antares, rises in the southeast for observers at northern latitudes by 9 p.m.

June 21–27
Clouds in space

If you're an avid skywatcher, no doubt you've seen many beautiful displays of natural skylight: halos, rainbows, sundogs, and crepuscular rays are a few examples. For the most part, however, these are unpredictable and can occur throughout the year. Only one can be said to be both predictable and rare: noctilucent clouds.

Noctilucent clouds (once referred to as luminous clouds) are cirrus-like clouds that form in the mesopause of our atmosphere at the very edge of space (between 75 and 90 kilometers). At this altitude, which is some five times higher than most of Earth's weather-making systems, temperatures may fall to $-100\,°C$ ($-150\,°F$) at the North Pole in summer. These extremely cold conditions allow the slight amount of water vapor in the mesopause to condense onto meteoric dust or, more likely, electrically charged atoms and molecules. These ice-encrusted particles are thought to be the source of the clouds.

The feathery stratum of noctilucent clouds displays iridescent hues, like the inside of an abalone shell, and exhibit a range of structure from veils and bands to waves and whirls. Noctilucent clouds are often tenuous enough for bright stars to shine through, making them a subtle phenomenon, though perhaps not as delicate as the zodiacal light (March 8 – 14).

To see noctilucent clouds, you need to be in the right place at the right time. The right time (for those in the Northern Hemisphere) is June and early July. The right place is typically between latitudes 45° and 60°. (They are also visible from these same latitudes in the south during midsummer in the Southern Hemisphere.) Most noctilucent clouds appear low in the twilight sky, usually not more than 10° above the horizon. They first become visible at twilight, when the Sun is 6° below the horizon and the contrast between the darkening sky and the brightening clouds is perceptible. Once the Sun sinks more than 16° below the horizon, it can no longer illuminate the clouds and they vanish.

Around the time of the summer solstice, however, when the Sun is never far below the horizon at high latitudes, a robust noctilucent display may be seen up to the zenith and even into the other half of the sky. This is especially true at latitudes higher than 60°, where the Sun never gets more than 6° below the horizon this time of year.

Like aurorae, noctilucent clouds are more often reported in the Northern Hemisphere, because the population density is greater at the high latitudes there than it is in the Southern Hemisphere. The countries in which observers most often report noctilucent clouds are Canada, the British Isles, Scandinavia, and Russia. In the United States, they may be seen in Alaska, although they have from time to time been seen in New York state.

Interestingly, reports of noctilucent clouds have doubled since the 1950s. Although the increase may be due to greater public awareness of the phenomenon, another explanation invokes an environmental concern. The proliferation of industry and the automobile during the twentieth century has resulted in widespread pollution. Some environmentalists think that an increasing number of hydrocarbons and other pollutants in the upper atmosphere may be supplementing the natural condensing nuclei at high altitudes. If true, the luminous beauty of noctilucent clouds may have a decidedly darker lining.

Also this week:
- The Summer Triangle, consisting of Vega, Deneb, and Altair, has fully risen as darkness falls this week. (See July 26 – August 1.)

June 28 – July 4
The glorious Milky Way

It's occasionally mistaken for a high, thin cloud when seen for the first time by city dwellers who don't often get a chance to see stars in a light-free sky. And in a sense, it *is* a cloud, a great cloud of stars vaulting across the sky like some ethereal, luminous bridge. More accurately, though, it is a galaxy – our galaxy – seen from the inside out.

We call it the Milky Way, a name derived from Roman mythology, which explained the whitish glow as a stream of milk from the breast of the goddess Hera, wife of Zeus.

According to the legend, Zeus was known for frequent philandering with mortal women. One marathon three-day liaison with the exceedingly comely young earth woman Alcmene (whom Zeus deceived into thinking he was her husband), resulted in the conception of Heracles, better known as Hercules the Hero. Zeus, like any immortal father, wanted his son to be immortal, too. But for that to happen, the baby Hercules first had to suckle the breast of Hera. Upon learning of Zeus's infidelity, however, Hera was understandably in no mood to cooperate with such a request. Nevertheless, Zeus lay Hercules upon the breast of the goddess as she slept. She soon awoke with a start and brushed the child from her. But the milk had already begun flowing and spurted across the heavens, forming the Milky Way.

Other legends about the Milky Way abound. American Indians believed it to be the celestial path taken by the souls of the dead on their arduous journey to Heaven, and that the brighter stars along the Milky Way were the campfires where they briefly stopped to rest. In China and Japan, the band of light was viewed as a river inhabited by spirits swimming toward the 'Land of Peaches.' The river metaphor appears again in Egypt, where the Milky Way was analogous to the Nile River, and in India where it represented the venerable Ganges. The Greeks viewed the Milky Way as the river Styx, across which the souls of the dead were ferried to the underworld.

Ancient astronomers thought the Milky Way was a kind of vapor emitted by the Earth. But Galileo put this notion to rest the first time he pointed his telescope at it and discovered that, in fact, the Milky Way consists of countless stars.

Today, you can see what Galileo saw using a small telescope or, better still, a pair of modest binoculars. The best conditions in which to see the Milky Way are under a dark sky. But even from the outskirts of a city, the cloudy band can be detected, especially when it is high overhead, which it is this week around 1:30 a.m. It can be seen earlier in the evening in a dark, moonless sky, arching above the eastern and southeastern horizons around 10:30 p.m.

Looking above the eastern horizon, you can easily see Cygnus the Swan, or the Northern Cross, lying on its side with the bright star Deneb shining at the head of the cross. This constellation runs down the middle of the Milky Way, the long axis of the cross pointing toward Aquila, Ophiuchus, and the star-rich southern realms of Sagittarius and Scorpius.

South of Cygnus, if your sky is dark enough, you may notice a dark cleft dividing the Milky Way, and that, indeed, just south of the tail of Aquila the Eagle, part of the Milky Way veers away to the southwest and dies out entirely. We see a major dark region in the southern sky, starting just past the tail of Scorpius and ending near Alpha Centauri. These dark 'rifts' are caused by concentrations of dense, interstellar dust lying along our line of sight. The dust is so thick that it curtains the stars behind it. If these dark clouds were not present, we'd see many more brilliant stars.

When you turn your binoculars on Sagittarius, which will be higher in the sky two weeks hence, you will be able to see many dark voids interspersed among the bright star clouds of that constellation (July 12 – 18). The most well-known 'hole' in the Milky Way appears in the Southern Hemisphere. The irregular patch of dark dust is so distinct to the unaided eye that it is called the Coal Sack.

As the night wears on, the Milky Way hoops up into the sky, revealing itself in all its glory. If your sky is dark enough, you may feel a little vertigo, for no other view gives you the sense of standing on the edge of the Earth and looking into a bottomless universe.

July 5 – 11
Galactic central

Northwest of the constellation Sagittarius, and north-northeast of the tail of Scorpius the Scorpion is a patch of special sky. What makes this region so unique is that it lies in the direction of the center of our galaxy. If you turn a telescope on this region you'll see a rich field of stars and several open clusters, but little nebulosity to speak of, and certainly no blazing central cluster of stars worthy of being called 'galactic central.'

Stepping back and taking in the entire Sagittarius/Scorpius region, we see a study in contrasts. Scorpius lies over on its side, with Sagittarius, in the form of the celestial 'teapot,' following closely behind. Out of the spout of the teapot issue great clouds of 'steam,' in other words, the billowy star clouds that so attract the eye. But just west of this area, the Milky Way's luminous appearance abruptly gives way to a rather dark chasm. This dark region consists of interposed lanes of dense interstellar dust – the raw material of the stars themselves.

Binoculars show that this dark zone isn't really that barren of stars. In fact, many stars are distributed evenly throughout, but overall, this section of the Milky Way is fainter than in Sagittarius.

What are we seeing? Where exactly is the center of our galaxy? To be precise, the direction toward the center is four and a half degrees west-northwest of Gamma (γ) Sagittarii on the Sagittarius–Ophiuchus border. But that is far from being the whole story.

To understand what we are seeing, we need to learn a few general facts about galaxies. The Milky Way is a type of galaxy called a spiral. Spiral galaxies are vast, disk-shaped systems of hundreds of billions of stars, which comprise the largest, most symmetrically shaped discrete structures in the universe. The gangly 'arms' of spiral galaxies consist of hot stars, clusters, and glowing nebulae that curve out from a bright central hub or bulge, giving the galaxy the general appearance of a whirl-pool or pinwheel.

From Earth's position, located in the middle of the disk far from the center of the Galaxy, a spiral arm appears as a great band of diffuse light arching across the heavens. When we view the Milky Way in Sagittarius and Scorpius, we're seeing one of the prominent spiral arms lying between us and the center of the Galaxy. This spiral arm is known as the Sagittarius arm. (To find out what we see when our view is perpendicu lar to the galactic plane, see May 31 – June 6.)

Although it's studded with star clusters and glowing nebulae, the Sagittarius arm is also curtained with much obscuring dust, the kind that you can see a little further west and north of Sagittarius. This dust prevents us from looking through the Sagittarius arm at the great galac-tic center itself.

The fact that this awe-inspiring sight is denied us may be somewhat disappointing. After all, if the Milky Way is as bright as it is, the center of our galaxy, packed with billions of stars, ought to be positively dazzling. Surely it would light up the night.

Well, yes and no. Astronomers estimate that the total luminosity of the central 10 light-years of our galaxy is on the order of 10 million or more suns. That sounds pretty bright, until you move it 27,000 light-years away, which is about the distance to galactic central. At that dis-tance, even a region of space as generous as 30 or 40 light-years across would at most cover a spot in our sky only 1 arcminute in apparent diameter (one-thirtieth the size of the full Moon). Thus, if interstellar dust wasn't a problem, our unaided eyes would see a central source occupying an apparent area of sky no bigger than Venus and no brighter than one of the stars in the cup of Big Dipper.

Okay, so much for the core itself. But interstellar dust obscures more than just the central few light-years of the Galaxy, it also dampens the light of billions of stars lying in the foreground and on either side of the central region. So, if we could somehow remove all the intervening dust, the combined light of all those stars converging on the galactic center would probably exceed that of the full Moon. In fact, you wouldn't be able to see much else in the summer sky except the combined light of billions upon billions of stars.

Of course, the Milky Way is rotating, and we along with it. At the Sun's distance from galactic central, the velocity amounts to about 220 kilometers a second, enabling the Sun to complete one 'orbit' every 230 million years. Given this rotation, plus the slight 'bobbing' motion our Sun makes as it moves through the disk, in another 10 million years or so, who knows what our view toward the center of the Galaxy may be like? Perhaps it will open up a bit, revealing more bright stars and clusters. Until then, we'll just have to be content with the beautiful contrasts, glowing star clouds, nebulae, and thousands of interesting deep-sky objects visible with binoculars or a small telescope.

July 12 – 18
Deep-sky treasures

This week around 10 o'clock, the summer Milky Way dominates the eastern half of the sky in the Northern Hemisphere. As the southern extent in Scorpius crosses the meridian, the northern realm in Cygnus is making its way toward the zenith. In a dark, moonless sky, a quick scan along the sky's 'spine' with binoculars shows such a throng of star clouds, clusters, and dark regions that the two-dimensional view you normally get when looking at the stars gives way to an almost three-dimensional feel. Among all this beauty are a number of deep-sky treasures that are easy to find in binoculars or a small telescope from a mostly light-free location.

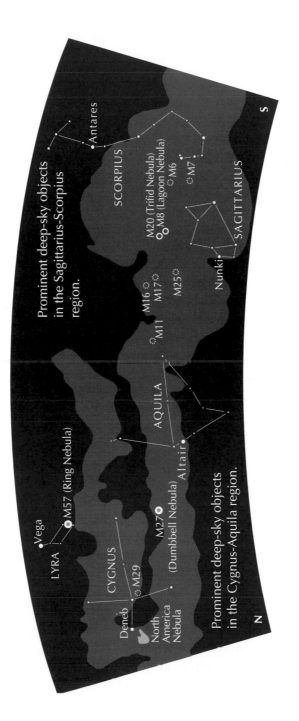

Prominent deep-sky objects in the Sagittarius–Scorpius region.

Prominent deep-sky objects in the Cygnus–Aquila region.

S

N

Antares

SCORPIUS

M20 (Trifid Nebula)
M8 (Lagoon Nebula)
M6
M7

SAGITTARIUS

Nunki

M16
M17
M25

M11

AQUILA

Altair

M27
(Dumbbell Nebula)

CYGNUS

M29

Deneb

North
America
Nebula

LYRA

Vega

M57 (Ring Nebula)

Beginning in Cygnus, just east of Deneb, your unaided eyes may sense a region of slightly enhanced brightness. Binoculars, which help increase the contrast between subtle glows and their darker surrounds, should reveal a wedge-shaped cloud of diffuse light. This is the renowned North America Nebula, a star-forming region with an uncanny resemblance to the continent. In fact, it is even oriented so that its geographic north and south directions are coincident with north and south in the sky, although geographic 'west' is celestial east and geographic 'east' is celestial west. Being almost 2° across, it is most definitely a binocular object since it would stretch out of the narrow field of view of a telescope. Look for the prominent dark patch on the nebula's western side – the 'Gulf of Mexico' – to orient yourself.

A careful sweep from Deneb through the middle of the Swan reveals a number of star clouds, particularly between Gamma (γ) Cygni and Albireo, the Swan's head. Located just southeast of Gamma is a knot of stars comprising the open cluster M29. Although not striking in the number of stellar members, M29 might be considered a modest challenge to locate in a binocular field full of stars. As small as it appears, however, this cluster may harbor more members than we can see. Areas of dark nebulosity could be concealing a great many more stars in M29, and we can only imagine how much brighter it would be if we could see all of its stars free of interstellar dust.

Continuing southward, carefully look about 8° east of Albireo in the constellation Vulpecula for a 'woolly' star. This object is known as the Dumbbell Nebula, although, to others, it resembles an apple core. It's classified as a planetary nebula, which actually has nothing whatsoever to do with planets. Many of them simply look small and disk-like in a telescope, like planets. The Dumbbell, however, is quite large and bears further scrutiny in a small telescope. Planetary nebulae are the outer envelopes of gas exhaled by dying stars. The Dumbbell is one of only 1,500 known in the Galaxy.

A sweep through Aquila and further southward reveals a more uniform distribution of stars as well as many small, distinctive star patterns or 'asterisms.' There are a number of planetary nebulae in this part of the sky, but, unfortunately, nothing bright enough to be seen in binoculars. Note, too, the dark rift west of Aquila. This is one of the more prominent lanes of dark interstellar dust cleaving the Milky Way.

As we enter the borders of eastern Serpens, Scutum, and the northern reaches of Sagittarius, we are besieged by deep-sky objects. The entire region brims with clusters and nebulae. M11, in northern Scutum, appears as a nebulous patch in binoculars but is, in fact, a compact open cluster. Further south lies M16, which is a combination star cluster and nebula. Small telescopes resolve it into a handful of bright blue stars. Larger telescopes, however, bring out the faint nebulosity enshrouding the group.

Moving into Sagittarius, you may notice a bar-shaped smudge of light just below M16. This is M17, known by its more popular names, the Omega, Horseshoe, or Swan nebula. And just a little further south and east is M25, one of the brightest, coarsest open clusters in the area.

On the western border of Sagittarius, just above the teapot 'spout,' binoculars will help you plainly see what the naked eye could only suggest. Glowing brightly against effervescing star clouds are two nebulae that are without a doubt the gems of the Milky Way, at least for observers in the Northern Hemisphere: the Trifid and the Lagoon nebulae. The Trifid lies north of the Lagoon, and in binoculars exhibits a ruddy glow that, upon closer inspection with a telescope, is subdivided by three lanes of dark obscuring dust. The Lagoon is the larger and brighter of the two. A small telescope boosts the contrast enough to show that it is bisected by a broad dust lane, from which the nebula gets its name.

The lowly binocular is too often given short shrift by amateur astronomers bent on using only telescopes to explore the sky. True, a large telescope can ferret out incredible detail and beauty in deep-sky objects, but a decent pair of 7×50 or 8×56 binoculars used beneath a dark, moonless sky provides the observer with an excellent means of stepping back and discovering the Milky Way's big-picture appeal. Besides, as any art student will tell you, if you spend all of your time examining the brush strokes, you can't really appreciate the scope of the painting.

July 19–25
Dog Days

For those of us living in the warmer temperate and tropic zones of the Northern Hemisphere, summer has taken hold and we are in full sweat.

The middle of the day is hot and still. There is no breeze, no benevolent cloud, and no change in the forecast. The air is full of buzzing cicadas, while on shaded, cool concrete porches dogs and cats sleep heavily, oblivious to passersby foolish enough to be out in the heat.

These are the 'Dog Days' of summer – hot, still, lazy days in which the summer Sun seems especially intense. Of course, nowadays, we attribute the heat to the fact that at this time of year the Northern Hemisphere receives more direct rays from the Sun. In ancient times, though, the Dog Days were associated with the bright star Sirius in the constellation Canis Major, the Greater Dog. Ironically, this star is primarily known for its appearance in the winter sky. In fact, we typically see it highest in the south in mid-February evenings.

You can see Sirius this week just at sunrise above the southeastern horizon. A line drawn through the belt stars of Orion and extended 'downward' leads to Sirius, the brightest and one of the nearest stars in all the night sky.

The name Sirius comes from the Greek *seir*, to shine, but is often translated to mean 'sparkling' or 'scorching.' Indeed, when it is just rising, it is a brilliant multicolored luminary, twinkling frantically in the atmosphere. The Egyptians venerated Sirius because its appearance in the east in the morning sky preceded the rising of the Nile River each year. Predicting this event was vital to the people living along the Nile's banks who needed to know when to store food, protect their supplies, and evacuate areas inundated by floodwaters. The star's appearance served as a warning, just as a barking dog warns his master of danger. Thus, Sirius came to be considered a watchdog. Its hieroglyph, a dog, often appears on temple walls throughout the Nile country.

This time of year, Sirius rises literally with the Sun and is hence outshined. But in just a couple of weeks, it will rise just before the Sun does, when it can briefly be seen. This is known as a helical rising. Recognizing this coincidence, and how intensely bright Sirius is, the Greeks came to associate the unusually hot days of summer with this star. Even in ancient times, Sirius was always within 10° to 15° of the meridian around local noon, and its rays were thought to mingle with the Sun's, thus compounding the heat. This week, if we could turn off the Sun at will, we would see Sirius very near the meridian at local noon, daylight saving time.

Sirius doesn't really contribute to the summer's heat; it's just a dazzlingly bright star. And although the Egyptians no longer require Sirius's services for flood control (they have the Aswan Dam now) the star still serves a useful purpose in warning us of the onslaught of summer and the many hot days yet to come.

Also this week:

- An elegant arc of stars known as Corona Borealis, the Northern Crown, is overhead for observers in the Northern Hemisphere as darkness falls. Look for it between Arcturus and Vega, but closer to Vega.
- The tail of Scorpius is high overhead by 9:30 p.m. for observers in the Southern Hemisphere.

July 26–August 1
The Summer Triangle

Go out tonight as dusk settles and look halfway up into the eastern sky. As twilight fades, you should be able to see three bright stars arranged in a large triangle. This venerable star pattern, known to amateur astronomers as the Summer Triangle, has long been the starting place for people wanting to learn the constellations.

Highest in the sky around 9 o'clock is Vega, a blue-white star whose name means 'swooping eagle.' Vega is not only the brightest luminary in the Triangle, it is also the fifth-brightest star in the sky (not counting the Sun). Vega is the brightest member of the constellation Lyra the Lyre – the harp of Orpheus. You can probably just see the other five stars of the group that make a triangle at one corner of a parallelogram.

Northeast of Vega is Deneb, which is Arabic for 'tail.' Since its name means tail, it must represent the tail of something – in this case, Cygnus the Swan. According to Roman mythology, this great white bird was sacred to Venus. Cygnus is one of the more distinctive summer constellations. Many prefer to call it the Northern Cross, a seventeenth century Christianized version of the star pattern. In this form, it represents the cross upon which Jesus was crucified, and that was later recovered by Saint Helena (mother of Constantine). Christians see a divine sign in the Northern Cross, for around 9 o'clock on Christmas Eve, the cross stands erect above the northwestern horizon just before setting.

Looking east at the rising Summer Triangle, July 27, 9:30 p.m.

As the cross, Deneb forms the head. But as a long-necked Swan, Deneb brings up the rear. The Swan is flying south directly through the Milky Way (which you can clearly see if you don't live near a large light-polluted city). The Swan's head is marked by the second-brightest star in Cygnus, Alberio. This star is worth viewing in a small telescope. Alberio itself has a decided orange hue, but very near it lies a smaller blue star that is a companion sun. The two strike a beautiful contrast.

Lying further south and forming the apex of the Summer Triangle is bright white Altair, the leading light of Aquila the Eagle. This celestial bird, however, is winging its way in a more northerly direction, opposite that of Cygnus. In nearly all ages, Aquila has been known as a bird of prey. The Arabic name for this constellation, *al-Nasr al-Tair*, the Flying Vulture, is where we get the name of its brightest star.

The Summer Triangle can be used as a guide to locate other constellations. A line extended from Deneb to Vega points to Hercules, while a line from Altair through Vega points at the squarish head of Draco the Dragon, a constellation that arches around the pole star. A line from Altair through Deneb leads to Cepheus the King, a faint pattern in the shape of a crude drawing of a house. And a line from Vega through Deneb shows the way to Andromeda and the Great Square of Pegasus, both of which rise late on midsummer evenings.

The Summer Triangle remains visible in the evening skies until late November, when it stands over the west-northwest horizon at sunset – with Deneb marking the apex and Altair and Vega forming the base.

Also this week:
- The Delta Aquarid meteor shower is active between July 29 and August 6. More meteors will be seen after 3 a.m. coming from the south. Typical numbers per hour range between 10 and 20.

August 2–8
A false comet

During the Comet Halley craze in 1985 and 1986, people were reporting sightings of 'new' comets almost daily. Of course, at the time there were probably more telescopes and binoculars pointing up at the sky from more backyards than at any previous period of civilization. Comet Halley had elevated all comets to Holy Grail status, so it's no wonder people continued looking for these fuzzy objects long after it had receded into the dark depths of the solar system.

Around July of 1986, such a fuzzy star was reportedly seen by many people high overhead in the constellation Hercules. It was bright enough to be seen in binoculars from the suburbs, and from a dark-sky site, it was nearly visible to the naked eye. Was this another comet? Unfortunately, no. What observers had 'discovered' was the great globular cluster in Hercules, cataloged as M13.

Globular clusters are compact balls of stars. A typical globular is about 150 light-years across and contains hundreds of thousands of stars. In long exposure photographs, they are spectacular objects,

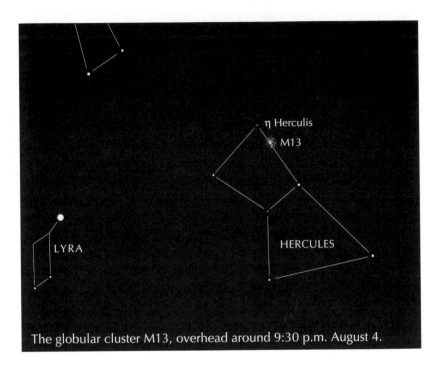

η Herculis

M13

LYRA

HERCULES

The globular cluster M13, overhead around 9:30 p.m. August 4.

looking a bit like a dense collection of swarming fireflies or an effervescing seltzer tablet. Despite their crowded appearance, however, globulars still consist mostly of empty space. At the center of M13, for example, there is, on average, one star for every cubic light-year. Nevertheless, the view from a planet near the center of one of these clusters would be magnificent. The sky would be filled with stars with the brightness of Venus and the full Moon.

To find this beautiful 'false comet,' go outside around 9:30 p.m. and look straight overhead for Hercules. This is a rather faint, sprawling constellation, but it has one distinctive characteristic: a wedge-shaped quadrilateral of stars at its center called the Keystone. Using binoculars, look along the western edge of this quadrilateral, but especially just south of its northwesternmost star, Eta (η) Herculis. Your eye should catch on a star that looks as if you're seeing it through a slight haze. That object is the great globular itself. A close inspection with a telescope, especially one located at a dark-sky site away from the city, reveals a granular, spherically-shaped aggregate of stars.

M13, like other globulars, lies very far from Earth. In fact, globulars comprise some of the most distant objects visible in our galaxy. The Hercules cluster is estimated to be about 25,000 light-years away, and is located, along with other globulars, in a halo that surrounds the central hub of our galaxy. The fuzzy glow you see tonight, which is the total light emitted by its million or so stars, left the cluster during Earth's last ice age.

You might also be interested in knowing that you are looking at one of the oldest objects in the universe. Astronomers estimate that the Hercules cluster may have formed between 10 billion and 15 billion years ago, just a few billion years after the universe came into being. These objects are nearly as old as time itself.

Meteor shower alert

While looking at the Hercules cluster, you may notice a few more meteors in the sky than usual. This week leads up to the peak of the Perseid meteor shower on the morning of August 12, and some of the meteors you see may be from that display, especially if they appear to be coming out of the northeast. This shower is active between July 17 and August 24, but the week before maximum should see a significant increase in the number of Perseid meteors. The best time to look for the Perseids is in the predawn hours, but it is not unusual for these meteors to be seen during the evening, particularly near the peak. See the next chapter for details.

Also this week:
- The bright ruddy star Antares is on the meridian after sunset in the Northern Hemisphere.
- In the Southern Hemisphere, Arcturus is low in the northwest by 9 p.m.

August 9–15
The Perseids: celestial fireworks

A bright shooting star is one of the most breathtaking sights in the night sky. What makes it all the more delightful is the fact that the observer

Perseid meteor radiant, after midnight, August 12.

happened to be looking in the right place at the right time when the unexpected event occurred.

Centuries ago, however, a meteoric display was an unwelcome sight, especially to people in power, because in most cases, a meteor was considered an ill omen that reflected on kings and military leaders. Sir James Frazer in his book *The Golden Bough* tells us that the reign of many ancient Greek kings was limited to eight years. The practice was founded on the belief that at the end of every eight-year period 'a new consecration, a fresh outpouring of the divine grace, was regarded as necessary in order to enable [kings] to discharge their civil and religious duties.' As part of the Spartan constitution, therefore, it was required that every eighth year, the ephors – the five Spartan magistrates who had power over the king – would choose a clear, moonless night to observe the sky.

If a meteor was seen during their vigil, they construed that the king had sinned against the deity, and suspended him from his functions.

Apparently, the practice lasted for centuries. Today few people would profess belief in the divinatory power of shooting stars. For those that do, however, the sky this week will be crisscrossed with celestial portents in the form of the Perseid meteor shower. This is probably not a good week to be king.

The after-midnight hours leading up to dawn on August 12 are usually the prime window to view the annual Perseid meteor shower. Hopefully, there won't be a bright Moon in the sky to interfere with the fainter meteors. But even if there is, you may still be able to see several of the the brighter ones flashing across the sky.

The Perseid shower is one of the most popular meteor displays of the year. One reason is because it occurs during high summer when people are more apt to be outside doing a little stargazing. The real reason for its popularity, however, is because of its activity. It was, in fact, once considered the finest of the annual meteor showers, although in recent years it has been edged out slightly by the Geminid shower, which occurs in December. (See December 6 – 12.) Nonetheless, the Perseids typically exhibit a large number of bright meteors, several of which may explode spectacularly or fragment and leave brief smoke trails (also known as smoke 'trains'). The brightest of the Perseids can be brighter than the brightest stars in the sky.

Given local sky conditions (the degree of light pollution and cloudiness) it's difficult to predict exactly how many meteors you may see. If there is no bright Moon in the sky at the time, you should be able to see between 20 and 30 meteors per hour in the predawn sky on the 12th. At that time, the radiant is overhead. If you trace the paths of these meteors backwards they should lead you to this location in the sky, which lies just northeast of the Double Cluster in the constellation Perseus.

The Perseids consist of the debris left over from Comet Swift–Tuttle, first seen in 1862. Astronomers determined that the comet had a period of 120 years, meaning that it should return to our part of the solar system every 120 years. Swift–Tuttle did not, however, return as expected in 1982, nor was it seen in the years following. In 1991, the Perseids put on a stronger showing than in previous years, leading some to speculate that the comet would return in 1992. Indeed, the comet passed near the Sun in December of that year and the August 1993 display was one of the most active of the century.

Don't be disappointed if your sky is cloudy the morning of the 12th. Because the shower is active over a period of about a month, the week before and the week after the peak should also be good for meteor watching. Go out in the early morning hours (4 a.m. to dawn may be the best time) and look overhead and toward the northeast. Use a reclining chair or sleeping bag, lie back, and catch a few falling stars.

August 16–22
The Native American Scorpion

Off in the south-southwest as evening twilight falls is a smoldering red star flickering like a lively flame. This is the star Antares, the heart of Scorpius the Scorpion. This southern constellation is one of the few star patterns that actually looks like what it is supposed to be. Three stars in a slightly curved row form the head, and from the middle star, a string of a dozen stars, including Antares, meanders southeast forming the J-shaped tail of the scorpion. The distinctive stinger, marked by the star Shaula, is embedded in one of the brightest and richest regions of the Milky Way. Train a pair of binoculars here and you will see star clouds and clusters galore.

Scorpius is the most southerly of the zodiacal constellations. It becomes visible each year in mid-April around 10 o'clock, first appearing as a T-shaped group rising in the southeast, with Antares, the scorpion's heart, shining dully through the thick atmosphere. By midnight, most of the entire constellation has cleared the trees, with Antares assuming its characteristic sanguine glow. The name Antares signifies 'Rival of Mars,' because of its striking resemblance to that planet's color.

Traditional mythology casts Scorpius in the role of the deadly arachnid that chased and finally stung Orion after the hunter boasted that he would rid the world of all animals. Note that the two constellations are placed on opposite sides of the sky, so that as Orion sets in the west, Scorpius rises in the east, forever in pursuit of the Hunter. The two can never be seen completely in the sky at the same time.

In Native American cultures, the stars we call Scorpius played a significant role in determining annual milestones. The Navajo referred to the upper body of Scorpius as 'First Big One.' When they saw this group rising, they knew spring would soon be over. The three stars forming the stinger of the scorpion were known as 'Rabbit Tracks,' because they look like the meandering tracks a rabbit leaves in the snow. The position of the Rabbit Tracks in the sky helped signal the beginning and end of the Navajo hunting season. When the open end of Rabbit Tracks tips toward Earth, as the tail does when the constellation dips toward the western horizon, it is fall, and hunting season begins. When the open end points upward, as it does upon rising in late spring, hunting season ends.

The Skidi Pawnee referred to the two stars comprising the scorpion's stinger – lambda (λ) and upsilon (υ) Scorpii – as the 'Swimming Ducks.' When the ducks first appeared just before dawn in late February, they knew that spring and the 'time of thunder' were close at hand. To the Skidi Pawnee Swimming Ducks represented loons that alerted the water birds to fly north again.

Today, most skywatchers associate the appearance of the scorpion with the onset of warm, fragrant nights and hot days. Still, there was a time long ago when some of the first people to inhabit North America depended on these star patterns to help them live. See if you can find First Big One, Rabbit Tracks, and Swimming Ducks this week.

Also this week:
• Teapot-shaped Sagittarius is nearly overhead by 9 p.m. in the Southern Hemisphere.

August 23–29
Summer holds on as autumn draws near

The bright summer stars Deneb, Altair, and Vega, which form the summer triangle (July 26 – August 1), still dominate the upper reaches of the sky this week. They will be with us at least two more months, though they trend a little more toward the west each night. It's still summer, but already a new crop of stars is rising in the east – the autumn sky – and by this time next month, they will reign supreme.

Most distinctive in this group is the Great Square of Pegasus, now nearly halfway up in the northeastern sky. It is indeed, a large squarish asterism, with a bright star marking each corner. You could place 26 full Moons from its eastern side to its western side. Since a full Moon is roughly half a degree in diameter, that means the Great Square is about 13 degrees wide.

The neck and head of the flying horse consist of a string of four stars stretching westward from the southwestern corner star, the brightest of the square's stars. The forelegs – two parallel strands of fainter stars – extend straight out from the square's northwestern star.

Appended to the square's northeastern corner star is another constellation: Andromeda the Maiden. Andromeda is still fairly low and not easy to make out, but by next month she will be easier to see. Incidentally, in this constellation lies the most distant object that can be seen with the unaided eye – the Andromeda Galaxy (November 1 – 7).

North and slightly east of the Great Square lies a distinctive zig-zag group of six stars shaped somewhat like an up-ended W. This is the constellation Cassiopeia the Queen. The grouping is supposed to depict the Queen's throne, which you can see this week if you stand on your head. The throne rises legs first. Cassiopeia lies in the Milky Way, so if you scan this constellation with binoculars you should have no trouble seeing many stars and a few star clusters. A beautiful 'double cluster' lies east of the northeasternmost star in Cassiopeia. Under a dark sky free of city light pollution, you can see this intriguing object without optical aid. It looks like a small cloudy patch of light, an enhancement of the Milky Way's glow. (See October 4 – 10.)

East of Cassiopeia, and still down among the trees, lies Perseus the Hero, a giraffe-shaped grouping of about 20 stars. (See, 'The rescue of Andromeda,' October 11 – 17.) This constellation is better placed around 11 o'clock this time of year. When you look in this direction of the sky, you are looking toward the outer disk of the Milky Way, exactly opposite from its center in Sagittarius.

It's still officially summer, and will be for another month. But the sky hints at the coming of autumn, of cooler days, and sometimes chilly nights. The seasons are indeed written in the stars.

August 30–September 5
The smile in the sky and the watery realm

Rising in the southeast this week around 7 o'clock, and well up by 8:30, is a faint wallflower of a constellation, but one that offers a big smile if you can find it. In fact, it is often referred to as the 'smile in the sky': Capricornus the Seagoat.

This constellation has come down through the ages unaltered in form. In the ancient star atlases, Capricornus was depicted as a goat with the tail of a fish. The constellation resides in a region once known as 'the Sea.' But more on that later.

According to astronomer-mythologist Elijah Burritt, Capricornus is associated with the god Pan. One day, while Pan was feasting with Jupiter and some other deities near the banks of the Nile River, the giant monster Typhon suddenly emerged from the underworld and attacked them, scattering the assemblage in all directions. In an attempt to hide themselves, each deity assumed a different shape. Pan, who had leapt into the river, turned his head and upper torso into a goat and his lower body, which was submerged, into a fish.

When Pan emerged from the river, he discovered that Typhon had torn Jupiter's legs and arms from his body. Pan blew such a loud, shrill note on his pipe that it frightened Typhon away. Then Pan retrieved Jupiter's scattered limbs and tied them back on his body, allowing the god to move once more. Jupiter then struck Typhon with one of his thunderbolts, sending the creature back to the underworld to lick his wounds.

The fourth century Latin grammarian Macrobius states that the Chaldeans, the ancient Semitic people of Babylonia, named the constellation the 'Wild Goat' because the Sun, after heading south for several months, always began its climb northward again when it reached this point in the sky. (Goats are known for their mountain-climbing abilities.) Indeed, Earth's Tropic of Capricorn region was so named because the southernmost point of the Sun's path used to be in this constellation. It has since shifted west into Sagittarius.

To find Capricornus, face southeast around 9:30 p.m. Look for Altair in Aquila high overhead, then locate Fomalhaut, the bright star in Piscis Austrinus, shining by itself near the southeastern horizon (September

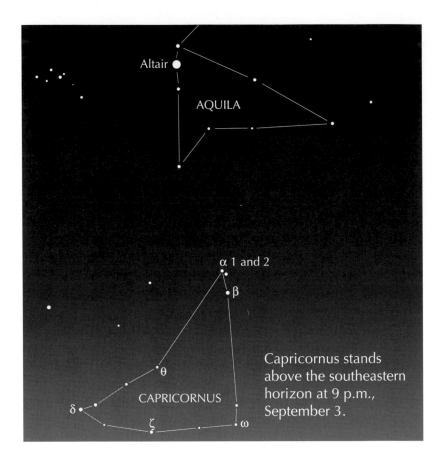

α 1 and 2

β

θ

δ

CAPRICORNUS

ζ

ω

Altair

AQUILA

Capricornus stands above the southeastern horizon at 9 p.m., September 3.

13 – 19). Capricornus lies just below a line drawn between these two stars.

This is a dim region of the sky, and even the brightest members of Capricornus hover only between magnitude 3 and 4. When traced out on the sky, they form a kind of Cheshire-cat smile. The brightest member, Deneb Algiedi (δ), marks the corner of the smile on the east, while Nu (ν) Capricorni, and Alpha (α) 1 (also known as Fredi) and Alpha (α) 2 Capricorni mark the western corner. Theta (θ), a hard-to-see magnitude 4.07 star, marks the middle of the smile's 'upper lip.' The lower half of the smile is outlined by Zeta (ζ) in the east, Omega (ω), in the middle, and, the constellation's second-brightest star, Beta (β), just below Alpha 1 and Alpha 2 on the western side of the smile.

Capricornus is the leading constellation of a group of beings that either inhabit the water or have some aqueous purpose. They are all well up in the sky by midnight and include Aquarius the Water Bearer, north and east of Capricornus; Piscis Austrinus the Southern Fish, south and east of same; Pisces the Fishes, south of the Great Square of Pegasus; and Cetus the Whale, south of Pisces. Sometimes included is the sinuous constellation Eridanus the River, which lies even further east nearer the winter constellations. Eridanus doesn't rise this time of year, however, until the wee hours, 2:30 or 3 a.m.

For observers equipped with telescopes, Capricornus contains only one deep-sky object of note, the small, dim globular cluster M30, which lies a few degrees east-southeast of Zeta (ζ) Capricorni. Just outside the constellation's border north of Theta Capricorni, you'll find M72, another modest globular, and the Saturn Nebula, a small but bright planetary nebula. In the same vicinity is M73, which is actually only an asterism of four or five stars. To the west, at the easternmost extreme of Sagittarius, is M75, a compact but fairly distinct globular cluster.

Also this week:
• Vega in Lyra is overhead for observers in the northern sky at dusk.

September 6–12
Time is distance among the stars

The stars shine with such a fine, sharp light, and often seem so close that people have a hard time grasping the fact that they really lie at incredible distances. Even the distance to the nearest star – the Sun – is mind boggling. At 93 million miles, the Sun is so distant that a photon moving at the speed of light – 186,000 miles a second – takes over eight minutes to reach Earth. On a more down-to-earth scale, say the legal highway speed of a typical automobile, the journey would take something like 300 years one way.

The imagination, though, can hardly comprehend the distances to the stars. In terms of miles, such physical spans are as abstract as the estimated number of grains of sand on the proverbial beach. For example, the nearest star to our Sun, the tiny red dwarf star Proxima

Centauri, is 25,200,000,000,000 miles away. This number is so large it's practically meaningless. If you've ever walked 20 miles, you have a pretty good idea how far that is. But 25 trillion? As essayist John McPhee writes in *Basin and Range*, 'Any number above a couple of thousand years – fifty thousand, fifty million – will with nearly equal effect awe the imagination to the point of paralysis.' Although McPhee was referring to geologic time, his comment applies equally well to astronomical distances, because distance is time on cosmic scales.

Astronomers combine units of distance and time into an intuitive term called the light-year. A single light-year is simply how far light travels in a year unimpeded through space. That distance is about 6 trillion miles. Dividing the distance to Proxima Centauri in miles by 6 trillion, you get 4.2. Hence, the distance to Proxima Centauri is 4.2 light-years. It is also its distance in time. In other words, light, traveling at a velocity of 186,000 miles per second, takes four and a quarter years to traverse the gulf of space between it and the Sun.

In the sky this week are a number of stars that, though they are nearly equal in brightness, are at radically different distances. At 9 o'clock, the Summer Triangle hangs suspended overhead like a giant starry mobile (July 26 – August 1). Deneb shines to the north, Vega, the brightest, to the west, and Aquila in the southeast. In actual distance, Altair is the nearest of these to the Sun, 16 light-years. Vega is next at 25 light-years. Deneb, however, lies 1,500 light-years away. To be as bright as it appears at that distance, Deneb must be an intrinsically bright star.

And indeed it is. Deneb is classified a supergiant. With a mass 25 times that of our Sun, it is also some 60,000 times brighter. If it could be transported to the distance of Vega, it would shine with an apparent magnitude of −7. For comparison, Sirius, the brightest star in the night sky is −1.4, and Venus, at its brightest, is about −4.8. A star of −7 would be over half as bright as the full Moon and would easily cast shadows.

If, on the other hand, Altair receded to Vega's distance, its brightness would dwindle to 2.3. This is slightly fainter than Polaris, the North Star. And if both Altair and Vega lay at Deneb's distance, they might not be visible to the naked eye at all.

Stars that lie within 10 or 20 light-years of the Sun are said to be in the 'solar neighborhood.' But stars that lie at Deneb's distance are considered remote. Nonetheless, even 1,500 light-years is nearby compared with the most distant stars we can see in our galaxy, which lie tens of thousands of light-years away, or tens of thousands of years in the past, if you want to read it that way.

More overwhelming still are the distances to galaxies and quasars – millions and billions of light-years. The most distant quasars detectable lie 11 to 12 billion light-years away, which dates back to a time when the universe was in its infancy. We are so removed from the 'now' of things in our galaxy and in the universe that we see remote objects not as they are, but as they were. Beyond the limited borders of Earth, the 'present' is an illusion stippled on the sky by stars shining in the dim past.

September 13–19
Fomalhaut: solitary star in the south

Late September skies are still rich with summer stars. Just after sunset, the northern Milky Way stretches across the top of the sky, with Deneb, Vega, and Altair huddled around the zenith, and Sagittarius crossing the meridian. In another month, however, these stars will slide off toward the west, and the autumnal constellations will begin taking their place, although without much fanfare from bright stars.

There is, however, one notable exception. Low in the southeast around 9 o'clock is a lone white beacon, a member of the otherwise dim constellation Piscis Austrinus the southern fish. The star's name is Fomalhaut, from the Arabic *Fum al Hut*, the fish's mouth. The appearance of the star this week signals the beginning of the end of summer.

Fomalhaut is one of the most southerly of all the prominent stars visible from latitude 40° N. In southern locales such as Chile or southern Australia, however, the star is nearly overhead this week by 11 o'clock. Because of the absence of nearby bright stars in this region of sky, Fomalhaut was an important star for ocean-going navigators.

To find Fomalhaut, you'll need a horizon with a clear, unobstructed view to the south and southeast. If your skies are hazy, use binoculars to help you penetrate the muck. Look just above treetop height for a solitary star, twinkling madly in the atmosphere. Its distance is only 22 light-years, meaning that the light you see from that star tonight left it 22 years ago.

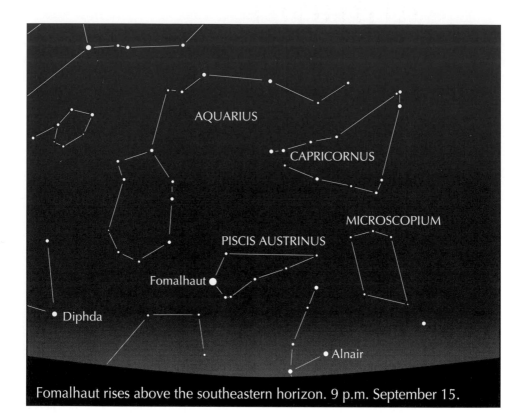

Fomalhaut rises above the southeastern horizon. 9 p.m. September 15.

Ostracized as it appears, Fomalhaut is nonetheless an important star. Radio images of the star show a doughnut-like structure of dust girdling this neighboring sun. The disk, which is roughly the size of our solar system, is cleared out in the middle, nearest the star, leading some astronomers to conclude that rocky planets may have already formed there. A similar dusty disk has also been observed orbiting Vega in Lyra (June 7 – 13) as well as stars in the southern constellations of Centaurus and Pictor. If these disks are indeed nascent solar systems, planets in our galaxy may be more common than previously believed. Unfortunately, like Vega, Fomalhaut is a hot star with an inherently short lifespan and it is likely it will die long before planets have formed.

Autumn

September 20–26
The Sun goes south

Those of you living in the northern climes know it already. The evenings are calm but edged with a decided chill, the smell of burning leaves permeates the air, and the Sun is setting earlier and earlier. Summer is definitely on the wane and autumn is at the door, where, if you can't yet feel it in the air, you can sense it in the changing sky.

The autumnal equinox, which marks the first day of autumn, occurs the moment the Sun crosses the celestial equator as it moves to the south. This usually occurs on September 23, though it can also occur a few hours earlier on September 22. A similar situation occurs around March 21, when the Sun heads north across the equator marking the beginning of spring. We call that event the vernal equinox (March 15 – 21).

Throughout the year, the Sun goes no farther south (or north) of the celestial equator than 23.3°, because that's how much Earth's axis is tilted with respect to the plane of the solar system. If you were standing at Earth's equator at noon on September 23, the Sun would appear directly overhead. But if you were located somewhere further north of the equator, the Sun would appear south of your the zenith at noon on that day by an amount equal to your latitude. Hence, the further north you are, the further south the Sun will appear. (In the Southern Hemisphere, the autumnal equinox marks the beginning of southern spring. There, the Sun gets higher in the sky as it advances southward in their part of the world, culminating with the beginning of southern summer around December 22.)

For example, on the day of the autumnal equinox at latitude 35° N, the

Sun will be 35° south of overhead. At latitude 45° N, the Sun will be 45° south of overhead. With the Sun located precisely over Earth's equator, almost every place on Earth with a clear horizon receives approximately 12 hours of sunshine and 12 hours of darkness. Equinox, incidentally, is Latin for 'equal nights.'

The fact that the Sun rises due east and sets due west at the equinox makes this a good time to pinpoint these cardinal directions from your home. Just go out at sunrise this week and note where the Sun rises in relation to nearby trees or buildings. Do the same thing at sundown. At noon, of course, north and south are perpendicular to the Sun's path across the sky.

Knowing where east, west, north, and south are from your observing location can help you orient yourself to the night sky, especially when you're using a star map or planisphere.

Also this week:
- The Northern Cross is on the meridian at dusk.
- Fomalhaut, brightest star in Piscis Austrinus, is overhead at 10 p.m. for observers in the Southern Hemisphere. (For observers in the Northern Hemisphere, see September 13 – 19.)

September 27 – October 3
Star time

It may officially be autumn, but the summer constellations are still hanging around overhead and in the west after sunset like actors reluctant to leave the stage. In fact, they seem frozen there. The starry sky seems no different than it did at this time last month. The reason: darkness falls earlier now with the Sun drifting ever southward, and this effect offsets the westward march of the constellations.

Nevertheless, the Earth keeps spinning around the Sun, and already a new crop of stars is rising in the east heralding not only a change in the sky but a change in seasons.

Low in the northeast around 8 o'clock, we see the Great Square of Pegasus suspended above the horizon, followed by Andromeda the Maiden, which is appended to the Square's northeasternmost star. Just to the south of the Square lies a circlet of stars marking the head of Pisces the Fishes. Cetus the Whale is beginning to make his appearance, too, though he won't be fully above the eastern horizon for another hour. Further south, dim Aquarius the Water Bearer is reaching out toward equally faint Capricornus, now nearing the meridian in the south.

Turning toward the northeast we can easily make out the zig-zag pattern of stars that compose the constellation Cassiopeia the Queen. Northeast of Cassiopeia is the giraffe-shaped constellation Perseus the Hero. And just rising, flickering violently on the horizon, is the bright star Capella in Auriga the Charioteer.

Three months from now at about this same time of night, the autumnal constellations will be roughly where the summer stars are now. The reason we see the constellations shift gradually westward throughout the year is because we measure each 'day' with respect to the Sun (a solar day), rather than to the stars (a sidereal day). A solar day differs slightly in length from a sidereal day, but in astronomy, slight differences at one end can amount to huge differences at the other.

If you do a little mental math, you can see that Earth rotates 360° in 24 hours. That amounts to a rate of 15° an hour, or 1° in 4 minutes. But by virtue of the Sun being larger in the sky than a distant point source like a star, and because the Earth moves in its orbit as it rotates, the solar day is slightly longer than the sidereal day. Our proximity to the Sun and Earth's orbital motion are the real culprits behind the longer solar day.

Suppose we begin at an arbitrary Day 1 with the upper limb of the Sun peeking over the horizon in the east. On Day 2, when Earth has completed one 360° rotation with respect to a distant star (essentially a point at infinity on the celestial sphere), it must still rotate a little more before the Sun can begin its breach of the horizon. Remember, the Sun is much closer than the stars. Just as the motion between a moving car and a fencepost by the highway at any one moment is more noticeable than the motion between the car and a distant mountain, the motion between Earth and the Sun is far more evident than the motion between Earth and the remote stars.

The additional amount the Earth must rotate so that the Sun assumes the same position in the sky it had the day before is 1/365th of a full turn, or nearly 1°. It takes Earth about 4 minutes to rotate this much, so the solar day is longer by nearly 4 minutes (3 minutes 56 seconds, to be exact). Hence, we see the stars rising about 4 minutes earlier each night. This adds up to 2 hours from one month to the next. So, if we see Capella rising this week around 8 o'clock it will rise at 6 o'clock a month from now, and by December 1 it will rise well before the Sun sets.

Imagine the chaos that would ensue if we measured our days by star time. Then it would be the Sun and our concept of day and night that would be radically altered. Let's use the star Sirius in Canis Major as an example. On November 1 Sirius crosses the meridian at about 4:30 a.m. If that equals '12 noon' Sirius time, then sunrise would occur 2 hours and 15 minutes later, at 2:15 p.m., Sirius time (6:45 a.m., Sun time). By February 1, Sirius is on the meridian at 12:30 a.m., Sun time. But if we take this as '12 noon,' then sunrise would occur around 7:30 p.m., Sirius time, rather than 7:30 a.m., conventional solar time.

Confusing? Yes, indeed. That's why the solar day works for civilization as we know it. The coming and going of the Sun is a lot more obvious than that of Sirius or any other star. Rather, the changing stars and seasons give us a different kind of time to dwell upon – not the brisk day-to-day flight of the Sun, but the more leisurely pace of changing months. In a sense, we may be more conscious of time when we're less distracted with daily events. Though the westward parade of stars happens incrementally, the changes tend to sneak up on us. When suddenly one evening we happen to notice a new group of constellations rising in the east, we can't help but realize with a start that, while the Sun has been marking time in days, the stars have been marking time in seasons.

October 4 – 10
The Double Cluster

Under the dry crisp skies of autumn, you should have no trouble spying a unique spray of stars lying in the winter Milky Way that is actually two star clusters in one. Facing northeast around 9 o'clock and looking halfway up into the sky you'll see Cassiopeia highest in the sky, with Perseus situated just below her. Between these two star groups – 7° southeast of Epsilon (ε) Cassiopeiae and 4° northwest of Eta (η) Persei – is a gauzy patch of light that is a slightly brighter portion of the Milky Way. Binoculars (7×50) reveal two clumps of stars lying in the part of the

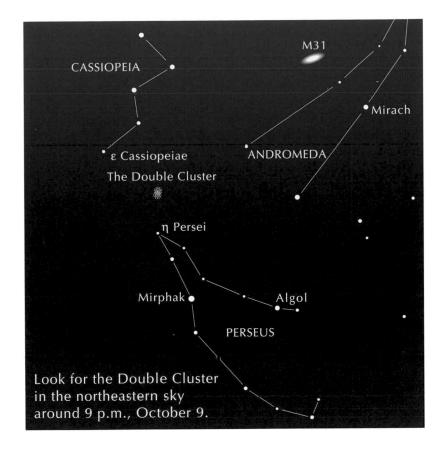

CASSIOPEIA

M31

Mirach

ε Cassiopeiae

ANDROMEDA

The Double Cluster

η Persei

Mirphak

Algol

PERSEUS

Look for the Double Cluster
in the northeastern sky
around 9 p.m., October 9.

Perseus figure known as the 'sword handle.' Each clump is a fine open cluster in its own right, but seen side by side in the sky, they make a stunning pair. The westernmost cluster is cataloged NGC 869 or 'h' Persei; the other is NGC 884 or Chi (χ) Persei. Collectively, this object is known as the Double Cluster.

It is well worth taking a look at these remarkable star clusters through a small telescope at low magnification so that both fit into the same field of view. Suddenly, the clumps resolve into dozens of individual strands of stars radiating from two concentrated groups. Astronomers have counted nearly 400 stars in the clusters, though there are undoubtedly many more than this shrouded in interstellar dust. Most of the visible stars are very hot and massive. Sprinkled along the outskirts of NGC 884, however, are a few red giant stars that can easily be picked out among the white and blue-white stars.

Astronomers think the Double Cluster first condensed out of interstellar dust and gas only a few million years ago, although the red giant stars associated with NGC 884 suggest that it is the older of the two. And though they appear to be next to each other in space, they are not. NGC 869 is about 7,000 light-years distant, and NGC 884 is approximately 1,200 light-years beyond that.

The Double Cluster is truly one of the night sky's most enthralling sights. Once you've seen it in binoculars or a telescope, you will eagerly return to it again and again.

Also this week:
- The Small Magellanic Cloud crosses the meridian at midnight in the southern sky. The Large Magellanic Cloud is easily visible 20° to the southeast. (See February 8 – 14.)

October 11 – 17
The rescue of Andromeda

Reigning over the crisp evening skies of mid-October are the signature constellations of autumn: the 'Great Square' of Pegasus the Winged Horse and Andromeda the Maiden, both of which are near the zenith; W-shaped Cassiopeia the Queen and giraffe-shaped Perseus the Hero lie a little further northeast. Just west of Cassiopeia, hovering around the celestial pole, is the faint constellation Cepheus the King, which is shaped like a child's drawing of a house. These are the main characters in a Greek mytho-drama, the moral of which describes the terrible wages of vanity.

Cassiopeia was the beautiful wife of Cepheus, the Ethiopian king of Jaffa* and the mother of the fair maiden Andromeda. Unfortunately, Cassiopeia was not only beautiful, she was incredibly vain. One day she made the mistake of openly boasting that she was more beautiful than even the sea nymphs. When word got around to the nymphs, they were,

* Also spelled Joppa. Formerly the biblical city of Iope, located northeast of Egypt. Since 1950, it's been a section of Tel Aviv.

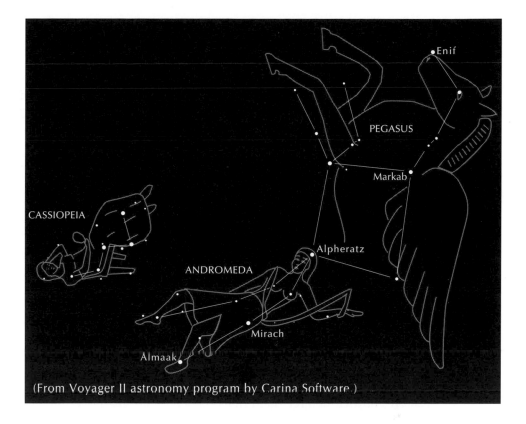

(From Voyager II astronomy program by Carina Software.)

needless to say, rather annoyed with Cassiopeia. They took their complaint to the sea god, Poseidon, who especially doted on his sea nymphs. Incensed, Poseidon sent the horrible sea creature, Cetus (in some myths, the sea monster is played by Draco the Dragon), to ravage the Ethiopian coast. During the latter part of October, this dim constellation can be found low in the southeast south of Pisces, well away from the main characters in this starry drama.

As soon as Cetus arrived in Jaffa, he proceeded to devour people and livestock and wreak havoc on the city. Panicking, King Cepheus consulted the powerful Oracle of Ammon. The oracle replied that the monster would be appeased only if the King sacrificed his cherished daughter Andromeda to the monster. Cepheus now found himself faced with the agonizing dilemma of having to choose between his daughter's life and the lives of his people. But as the death and mayhem continued,

Cepheus capitulated to the oracle's counsel and reluctantly ordered Andromeda chained to a rock near the crashing waves.

Cepheus and Cassiopeia were beside themselves with anguish and fear for Andromeda, an innocent victim of her parent's folly. They were standing near their unfortunate daughter, wailing and wringing their hands, when the sea creature appeared riding upon the waves. Just when things looked blackest for Andromeda, Perseus the Hero came upon the scene riding the winged horse Pegasus. He had come from slaying Medusa, one of the three Gorgons whose dreadful glance turned people, and monsters, into stone. (Pegasus, in fact, had sprung from Medusa's blood when it spilled upon the ground.)

Upon beholding the terrible scene, Perseus flew down and, stricken with Andromeda's beauty, offered to rescue her if the King granted him the maiden's hand in marriage. Having no other option, the King agreed. Perseus, who still had the head of the Gorgon, now held it up before the startled sea creature, which promptly turned into a stone. Perseus then gallantly swept up Andromeda on his trusty flying steed, and off they flew, presumably to live happily ever after.

Mythology being what it is (i.e., a series of fabulous tales told and retold), this is only one version of the rescue of Andromeda. In another rendering, for example, Perseus arrives on the scene on winged feet and sans the head of Medusa. He still negotiates his marriage to Andromeda with her father, but slays the creature with his sword in a gory struggle.

No matter how the story played out, the result was the same. The gods honored the participants in the drama by placing them in the sky, although 'honored' might not be the right word for Cassiopeia. For much of her passage across the sky, we see her sitting on her throne upside down. She's only upright when the constellation swings over to the western sky in spring, just before it sets. Ever at her side is Cepheus, brooding, no doubt, over why the oracle didn't suggest that his wife, rather than his daughter, be sacrificed to the sea monster. At least for Perseus, Cassiopeia's vanity paid off.

October 18–24
Shooting stars from Orion

If you can't sleep this week, try counting a few shooting stars. There will be a few more of them in the night sky than usual, especially during the wee hours of October 22. Go outside around 2 a.m. and look toward the southeast for faint, fleeting streaks of light among the stars. If you see one streaking out of the asterism forming Orion's upheld club, you'll have spotted an Orionid meteor.

The term 'meteor shower' is a misnomer. Meteors don't 'rain' down from the sky, and neither do they pelt the ground with meteorites. A meteor shower merely constitutes an increase in the number of meteors per hour that appear to emanate from a particular area of the sky, known as the radiant. In the case of the Orionids, if you track their streaks backwards across the sky, they all lead to a location in the constellation Orion the Hunter (hence the name, Orionids).

This week Orion rises in the southeast around midnight and is well up by 2 a.m. Look for its characteristic three stars in a straight row, which marks the hunter's belt. On opposite sides of the belt are two bright stars: brilliant blue-white star Rigel to the lower right or southwest, and ruddy Betelgeuse to the upper left, or northeast. The meteors appear to emerge from the hunter's club (or sword, if you prefer), which is formed by a scattering of faint stars arcing above Betelgeuse, below the feet of Gemini the Twins.

Like most meteor showers, the Orionids were spawned by ice, dust, and small stones spewed into space by a comet – Halley's Comet in this case – during its inbound journey to the Sun. The Eta Aquarids, seen in May, were also spawned by debris from Comet Halley (May 3 – 9).

You can expect to see between 10 and 30 Orionid meteors per hour on a dark moonless night. They're swift, faint, and may display color variations and persistent trains. One nice thing about the Orionids is that you can see them nearly all month. Unlike the more intense meteor showers that have sharp peaks and are spectacular for only one or two nights, the number of Orionids per hour climbs slowly from October 2 until its peak 20 days later. On days around the peak, several maxima regularly occur. After that, the numbers drop off slowly until around November 7. So, if you're vigilant, and prone to insomnia, you can expect to see a number of these shooting stars during the late night and early morning hours for the rest of the month.

October 25–31
Five southern birds

Call it a starry roost or celestial aviary, but this week after sunset, five constellations representing exotic or mythical birds flock around the south celestial pole.

Furthest north, and thus visible low on the southern horizon from mid-latitudes in the Northern Hemisphere, are Grus the Crane and Phoenix, the great bird that rose from the ashes of its predecessor. Three other birds lie closer still to the southern pole: Tucana the Toucan, Pavo the Peacock, and Apus the Bird of Paradise. Of these three, only Tucana can be glimpsed from latitude 30° in the Northern Hemisphere.

All five constellations were created by the German celestial cartographer Johann Bayer in 1603,* and only two are mentioned in classical mythology. The peacock figures prominently in the story of Argos, the builder of the ship, *Argo*, which was used by Jason and the Argonauts in their quest for the golden fleece. When Argos died, Hera changed him into a peacock and placed him in the southern sky not far from his ship.

Pavo is a fairly large and indistinct constellation. Its brightest star (which shines only slightly brighter than magnitude 2), is called the Peacock. The Peacock marks the bird's eye, while Gamma (γ), Beta (β), and Delta (δ) denote its breast and back. The plumage is set among the stars Lambda (λ), Nu (ν), Xi (ξ), Iota (ι), Pi (π), and Eta (η), all 4th- and 5th-magnitude stars.

The fabulous Phoenix was said to live for half a millennium. At the end of its life, it settled on a nest made of spices and fragrant leaves. When the Sun was next overhead, the rays ignited the nest, cremating the Phoenix in the flames. From the ashes, however, a new Phoenix arose, wrapped the remains of the dead Phoenix in myrrh, and burned them in the city of Heliopolis as an offering to the Sun. The cyclical metaphor can be traced to ancient China, Egypt, India, and Persia, where the

---

* In addition, Bayer created six other constellations: Hydrus the Water Snake, Dorado the Swordfish, Volans the Flying Fish, Indus the Indian, Chamaeleon the Chameleon, and Musca the Fly.

Phoenix was an emblem of cyclic patterns and immortality. Ankaa (magnitude 2.3), usually representing the eye of the Phoenix, can be seen just above the southern horizon at latitude 40° north this week around 9 o'clock.

Grus is the Latin name for crane, although to the Spaniards, the constellation of this name was considered a flamingo, which is a different family of bird entirely. Either way, the celestial version has the wing span and long neck of most wading water fowl. This bird's head is usually depicted by Gamma (γ) Gruis (magnitude 3). Its right wing stretches out to Iota (ι) (magnitude 4) while its left wing is marked by Grus's brightest star, Alnair, (magnitude 1.7), which can just be glimpsed from north latitude 30° southwest of Fomalhaut this time of year.

Southeast of Grus, the pentagonal form of Tucana spans the meridian this week around 7 o'clock. This is not, however, a prominent constellation by any stretch. Its brightest member, Alpha (α) Tucanae, is a little brighter than a magnitude 3 star. The Toucan is perched upon the Small Magellanic Cloud, its eye represented by the blue star Gamma (γ) Tucanae (magnitude 4), and its beak by Alpha. The southern bird's tail stretches nearly to the bright star Achernar in Eridanus the River.

Tucana is famous for harboring one of the most spectacular globular clusters in all the sky, 47 Tucanae. This object is bright enough to be seen with the unaided eye as a 4th-magnitude cottony patch of light, and, in fact, often appears plotted as a star in early star maps. It is second only to Omega (ω) Centauri in brightness. A 4- to 6-inch telescope reveals a brilliant, compact ball of tens of thousands of stars.

Just 11° west of the south celestial pole is humble Apus, a dogleg-shaped group of four main stars. Alpha (α), its brightest member, is only magnitude 3.8. Ironically, this unassuming asterism is supposed to represent the beautiful Bird of Paradise, any of the brilliantly colored plumed birds of the New Guinea area. Apus does have one asset, though: the majestic face-on spiral galaxy NGC 2997. Seen in a large backyard telescope, this galaxy is an elongated whirlpool of milky light. Its spiral arms and overall structure are thought to be similar to our own Milky Way galaxy.

Also this week:
• The bright southern star Fomalhaut can be seen due south in the Northern Hemisphere around 9 o'clock. (See September 13–19.)

November 1–7
A stepping stone to the universe

If you find yourself away from city light pollution and under a clear haze-free sky this week, look almost straight overhead around 10 o'clock.* Your eye may catch upon a faint glow, like a patch of thin cirrus, nestled between the stars of Andromeda and Cassiopeia. It will probably be something you see out of the corner of your eye, and you may wonder if you're seeing anything at all.

Indeed you are. What you're looking at is the Andromeda Galaxy, a system of several hundred billion stars that lies across a gulf of intergalactic space, 2¼ million light-years away. It is the nearest major galaxy to our Milky Way and ranks as one of the most distant objects visible to the unaided eye. The light arriving tonight from the Andromeda Galaxy was emitted over 2 million years ago, when early hominids like *Australopithecus* were fashioning rudimentary stone tools and establishing the first social order.

If you have binoculars handy (at least 7×50s) be sure to take a look at this object. Suddenly, what was a vague glow becomes a distinct oval patch of light with a bright center. If you could see the full extent of this galaxy in binoculars – as revealed by long exposure photographs – it would cover an area of sky more than twice that of the full Moon. Its most compelling features are its size, symmetry, and the dramatic way it stands out so starkly against the surrounding blackness.

Today, astronomers estimate there are billions of galaxies in the universe, each containing hundreds of billions of stars. During the early part of this century, however, such objects were thought to be gaseous disks residing in our galaxy. A great many exhibited striking whorl-like structures and were thus referred to as 'spiral nebulae.' Then in 1925, the American astronomer Edwin Hubble determined the distance to the spiral nebula in Andromeda by comparing the brightnesses of a class of nearby known stars called Cepheids to those in the Andromeda nebula.

Cepheids, named for the archetypal star of this class, Delta (δ) Cephei, fluctuate regularly in brightness. Like lighthouse beacons, they

* If you live in the Southern Hemisphere (Sydney, Australia, Cape Town, South Africa, about latitude 33°S) you must look about 15° above your north horizon at this time; at lesser southern latitudes (Lima, Peru, latitude 12°S), look anywhere between 30° and 40° above the north horizon.

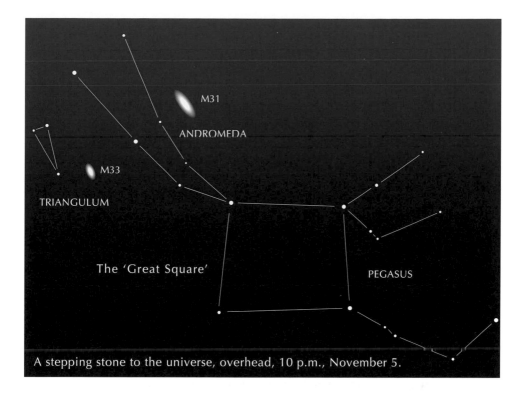

A stepping stone to the universe, overhead, 10 p.m., November 5.

can be used as 'standard candles' for estimating distance. Hubble discovered these same types of stars in the Andromeda nebula, as well as another spiral-type object, M33. By comparing the brightness of these stars to the ones in our galaxy, he showed that both the Andromeda nebula and M33 lay far beyond the boundaries of our galaxy.*

This epochal discovery forced astronomers to realize that the universe was vaster than had been imagined. For beyond even the nearest galaxies, themselves inconceivably remote, lay wave upon wave of galaxies stretching out to distances where, as Hubble himself once

---

* Many observers claim to be able to see M33 with the unaided eye. Located 14° southeast of M31 in the constellation Triangulum, this galaxy, known as the Pinwheel, is a bit larger on the sky than the apparent size of the full Moon, but its light is also spread out over this area, making it difficult to detect in binoculars, even in dark skies. Still, if you can see M33 with the naked eye, that would make this object the most distant one you can see without the aid of a telescope. Its distance is 2.5 million light-years.

remarked, 'we measure shadows and we search among ghostly errors of measurement for landmarks that are scarcely more substantial.'

Galaxies are the largest, coherent building blocks of the universe. Together they make up clusters and superclusters of galaxies, some of which stretch for hundreds of millions of light-years in length.

Closer to home, however, the nearest group of galaxies comprise a humble little cluster astronomers call the Local Group. In addition to M31 and M33, the Local Group includes the Large and Small Magellanic Clouds (February 8 – 14), and over two dozen other elliptical, irregular, and spiral galaxies, including our own Milky Way. Galaxies may appear isolated in the universe, but in fact they, like stars, are more gregarious than they seem.

November 8 – 14
The Leonid meteor shower

Most years, the Leonid meteor shower tends to be rather humdrum. At most, one can expect to see maybe 10 to 20 meteors per hour zip fleetingly across the sky in a dark location out in the country. But every 33 years, this shower suddenly becomes a storm of 'shooting stars.' The last major meteor storm generated by the Leonids occurred in 1966 when observers in the central United States saw over 5,000 meteors per hour.

Why does a meteor storm happen only occasionally? The answer doesn't have as much to do with where meteors are in space as much as it has to do with where comets are. Comets are the primary sources of meteoroids.

Comets are like snow-encrusted cars barreling down a highway in winter, leaving in their wake an icy, blustery trail. When a comet comes out of the deep freeze of space and visits the inner solar system, its surface is warmed by the Sun. Part of its icy crust suddenly turns to vapor, which solar radiation blows off the comet's surface, leaving a trail of debris through which Earth passes at a specific time of the year. This 'blowback' material contains ice, dust, and small gravelly material – stuff that later burns up in our atmosphere during a meteor shower.

The Leonids in particular are composed of debris from periodic Comet Tempel–Tuttle. This comet has a period of 33.18 years. As the

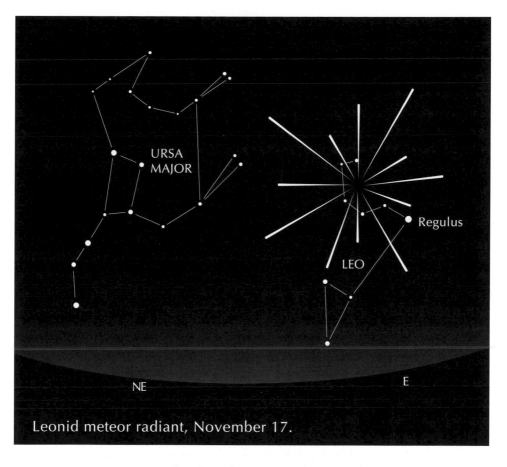

Leonid meteor radiant, November 17.

comet approaches the solar system, it brings with it a concentrated trail of debris, which Earth passes through. When this happens, we see a greater display of meteors than usual.

Look for the Leonids on two mornings around November 17. If there is a half Moon or less in the sky, the extra light shouldn't be too much of a problem. If the Moon is full, however, you'll probably only see the brightest of the Leonids. Leo rises in the east around 1:30 a.m., local time, just as Earth is meeting these meteors head on. The greatest number, however, will be seen when the radiant (near the Lion's 'sickle') is halfway up in the eastern sky. That occurs around 4 a.m.

In a typical year, you can expect to see a half-dozen or more meteors per hour from a dark-sky sight. Then again, if we pass through a 'knot' of debris, as occasionally happens with the Leonids, you may see a brief 'flurry' of meteors.

November 15–21
The Pleiades rise at sunset

Just before dawn in early June, a tiny dipper-shaped group of stars in Taurus can be seen shining faintly through the strong twilight. This event is the 'helical rising' of the Pleiades, the cluster's first appearance after coming from behind the Sun. For the rest of the year, the Pleiades rise earlier each day, mounting ever higher in the sky until this week, when they rise as the Sun sets and are in the sky all night. Look toward the east after sunset to see this distinctive nest of stars glimmering among the trees.

The annual appearance, culmination, and disappearance (helical setting) of the Pleiades was used to establish important religious and agricultural dates for many cultures, including the ancient Maya and Aztecs in Mexico, and the Pueblos in the southwestern U.S. The Navajo revered the Pleiades as father and creator of the world.

The name most frequently associated with this cluster is the Seven Sisters, although only six stars are generally visible to the naked eye. The story goes that one of the sisters married a mere mortal and for this reason shines less brightly than her sisters.

The cluster's appearance in early November reminds those in the southern U.S. of the coming of winter (cold weather has, no doubt, already arrived for those in northern climes). In binoculars, the Pleiades nexus appears like diamond chips on black velvet. The entire group is clumped within an apparent area about one and a half degrees – three times the diameter of the full Moon – so the high magnification and limited field of view of a telescope won't show you the entire setting of this group.

Astronomers have cataloged several hundred members scattered in and around the Pleiades. As star lives are measured, astronomers believe the cluster is comparatively young, a few hundred million years old. Recent research even suggests that some star birth is ongoing in the Pleiades.

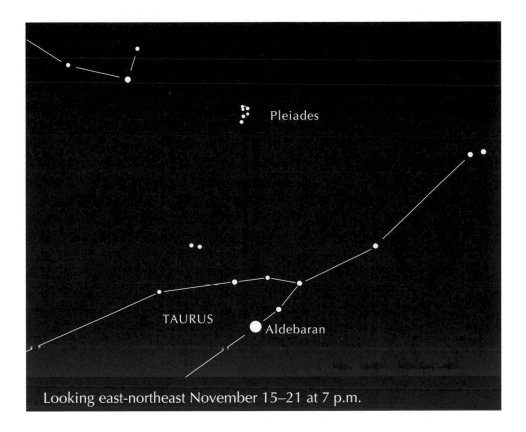

Pleiades

TAURUS
Aldebaran

Looking east-northeast November 15–21 at 7 p.m.

Today, people idly scanning the starry sky will pause upon the Pleiades and remark how lovely it looks. Little do they know they are looking at a celestial hallmark that was woven into the fabric of the lives of many ancient civilizations. As direct or indirect descendents of those civilizations, the Pleiades cluster appeals to that primal level in our brains that harkens back to a past still encoded in our genes.

November 22–28
A star that 'winks'

The stars may look unchanged from night to night, but in reality, they're changing constantly. We don't see the changes because, from our perspective, they are too subtle for the human eye. For example, astronomers know the stars are moving rapidly through space, but space is so vast and the stars so far away that we don't notice their motion except when we measure their positions with sensitive instruments over extended periods of time.

One change that is fairly noticeable occurs when a star increases or decreases its brightness significantly. This happens more often than you might think. In fact, there are literally hundreds of thousands of these so-called 'variable stars.' In most cases, it is difficult to see them change, but there are a few notable exceptions. One of these is the star Algol in the constellation Perseus, which is well up in the northeastern sky an hour or so after sunset this week.

In classical times, Algol represented the head of the Gorgon, which Perseus slew just before his rescue of Andromeda. The name Algol is, in fact, derived from the Arabic *Al Ra's al Ghul*, the 'demon's head.' It is speculated, however, that the star's odd variability – which went so against the deep-rooted belief that the heavens were immutable – may have been what inspired ancient astrologers to consider it the most unfortunate star (a *desastre*) in the heavens.

Normally, Algol is quite bright, almost as bright as Polaris, the North Star. But every 2 days, 20 hours, 48 minutes, and 56 seconds it fades away to being fainter than the faintest stars in the Big Dipper. You can easily watch these brightness changes with your eyes or binoculars over a period of several nights.

The regular brightness fluctuations result when a fainter companion star passes in front of Algol, partially blocking the brighter star's light from our view. This type of variable star is known as an eclipsing binary, of which Algol is the archetype.

To find Algol, look halfway up into the northeastern sky around 8 o'clock for the giraffe-shaped constellation Perseus. During maximum, Algol is the second-brightest star in that constellation, located in the westerly – and shorter – 'leg' of the giraffe. During its 10-hour minimum, it is slightly brighter than an adjacent star, Rho (ρ) Persei.

Also this week:
- Orion rises in the southeast by 9 p.m.
- In the Southern Hemisphere, the bright elongated spiral galaxy NGC 253, easily visible in binoculars, is overhead near the meridian between 8

and 8:30 p.m. Look for a long patch of milky light. Skies need to be fairly dark and haze-free.

November 29 – December 5
Betelgeuse, Betelgeux, Betelgeuze

Bordering the western edge of the glittering Milky Way, Orion's second-brightest star, Betelgeuse, is like a ruby set among a coal seam.

Astronomers know volumes about this compelling star. They know, for example, that it belongs to a class of elderly stars called red super-giants, which are approaching the end of their lives. As these stars die, their tenuous outer envelopes balloon out to enormous sizes. In the case of Betelgeuse, this is about 500 million miles across. Placed at the Sun's position, it would envelop all the inner planets and much of the asteroid belt; Jupiter would likely orbit within the star's outermost rarefied atmosphere.

Astronomers know Betelgeuse's surface temperature (about 2,100 kelvins, or 3,350 degrees Fahrenheit). They know its mass (20 times the Sun's mass), its luminosity (14,000 Suns), and its approximate distance (540 light-years). Observations of its light output show that it fluctuates noticeably in brightness (about 1 magnitude) and in size (as much as 170 million miles) every 5.8 years. Special imaging techniques have even imaged Betelgeuse's bloated surface, revealing that it is emblazoned with 'hot spots,' regions where the gas is brighter than its surroundings.

But there is still one searing question that has yet to be put to rest, one that plagues and bedevils both astronomers and the general public alike: how do you pronounce (and spell) 'Betelgeuse'?

It's a mouthful, no doubt about it. And it often elicits smiles and guffaws from the public when its name is mentioned at star parties. ('Beetle *what*?') Of the few astronomy guides that provide the phonetics for star names and constellations there seems no real consensus.

The Observer's Handbook, published by the Royal Astronomical Society of Canada, favors BET-el-jooz as the correct pronunciation. Martha Evans Martin, in her classic work *The Friendly Stars*, prefers BET-el-gerz. Richard Hinkley Allen in *Star Names, Their Lore and Meaning*, spells it Betelgeuze, which may be the source of Martin's pronunciation.

Allen, who is more concerned with astronomical etymology, avoids the pronunciation issue, but does present alternate spellings, these being Betelguese and Betelgeux.

Webster's lists it alternately as BET-l-jooz and BET-l-jœz (the œ being pronounced like the French 'eu,' as in *jeu d'esprit*). I've heard more than a few erudite amateur astronomers variously pronounce it BET-el-geeze, BET-el-gœz, and BET-el-gez. And, of course, we know what Hollywood did to it: BEETLE juice.

I suppose in some respects, the proper pronunciation of this star name is an academic issue that shouldn't much concern skywatchers. Besides, the etymon of Betelgeuse is the Arabic phrase *Ibt al Jauzah*, which means 'Armpit of the Central One.' Such a word, then, can't be considered elegant, no matter how you pronounce it. Therefore, I suggest you say it in such a way that doesn't set your teeth on edge or cause you to inadvertently spit in someone's eye. Just remember, it's a beautiful, old star; don't judge it by its pronunciation.

December 6–12
The flashy Geminid meteor shower

The after-midnight sky this week will be punctuated with shooting stars as the Geminid meteor shower approaches its peak on December 13. Observers may see as many as 50 bright meteors per hour, unless a bright Moon interferes.

Like all meteor showers, the Geminids are named for the constellation or star from which they appear to emanate. Gemini rises around 6:30 p.m. local time and so meteors may be seen in the sky all evening, most coming out of the northeast. Greater numbers of meteors, however, will be seen after midnight and in the morning hours as the night side of Earth faces into its orbit, sweeping up more cosmic debris. By 12:30 a.m., Gemini is directly overhead and this is the direction from which the meteors will emanate.

Typically, Geminid meteors are bright, but only a few leave smoke trails. They are predominantly white in color and can be seen almost anywhere in the sky. To be a true Geminid, as opposed to being a 'sporadic' meteor, its fiery path across the sky must lead back to the constel-

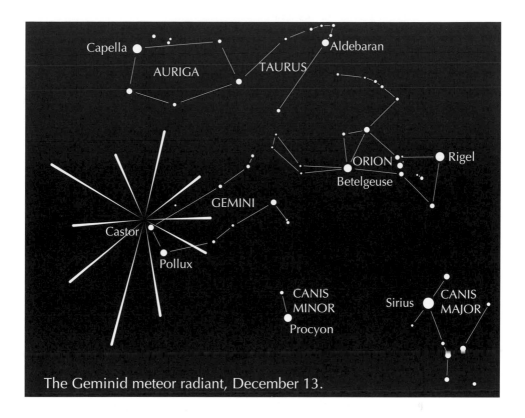

The Geminid meteor radiant, December 13.

lation Gemini the Twins. Gemini is a distinctive group, hosting two bright stars: white Castor and orange Pollux. These two lie side by side in the sky northeast of the great constellation Orion.

Unlike most other meteor streams, the Geminids are derived from an asteroid, designated 3200 Phaethon, which occasionally swings near Earth. Asteroids are not usually credited with creating the debris trails responsible for meteor showers, but they may be more responsible than previously thought. In 1997, observations of another asteroid, 253 Mathilde, by the Near Earth Asteroid Rendezvous (NEAR) spacecraft, confirmed that it had not only been heavily pummeled by meteors in the past, but that it is also composed of very porous and uniform material. Perhaps, then, particles might easily be blasted away from such a 'fluffy' asteroid, creating debris streams that could result in brief or even long-term meteor showers not unlike the Geminid display.

Also this week:

- In the Southern Hemisphere, the Puppid–Velid meteor shower is active. This shower exhibits several radiants, one in Puppis, one in Vela, and one in Carina. Moreover, it has several peaks. The best time to catch at least one peak may occur this week around December 9. The second occurs later this month on or about December 26. Expect between 5 and 15 meteors per hour. Puppis and Vela rise around 7:30 p.m., but more meteors may be seen when they are higher in the sky, after midnight.

December 13–20
Winter's eve

Who needs a calendar to tell you when the seasons are getting ready to change? Sometimes, all you need is a clear view of the stars.

You know that winter will soon turn to spring when you see the serpentine shape of Scorpius rising in the southeastern predawn sky, signaling the return of living things to the world. Spring is not far off when the Dipper stands on its handle in the northeastern evening sky, preparing to spill a cupful of rain over the land. Arcturus, fuming in the eastern evening sky after sunset, heralds the warmer days of summer, while the three stars of Orion the Hunter in the southeast around 4 a.m., declare the onset of autumn and the hunting season.

As a species, we are predisposed to seek out cycles that track through our lives like well-timed trains. By noticing the comings and goings of the stars, we not only reaffirm our link to the cosmos but acknowledge the legacy of our primal history when, as intelligence was dawning in our developing brains, we sought out order in a disordered universe.

As astronomy enthusiasts we are closer to this legacy perhaps than those for whom the heavens are an inscrutable and abstract muddle. If you're attuned to the ways of the sky, the diurnal motions of the Sun and Moon have a way of accumulating and replaying in the imagination day after day, year after year until one day, regarding the abbreviated shadows of midsummer or the rapid dimming of the day in autumn you hear yourself saying, 'I've seen this before.'

And so it goes, year after year, century after century. The surface of the

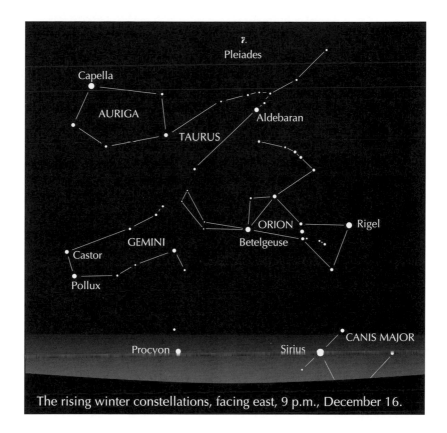

The rising winter constellations, facing east, 9 p.m., December 16.

Earth could be remade many times over before the stars would veer noticeably from their courses. Just as it did thousands of years ago at this time of year, golden-hued Capella, brightest star in Auriga the Charioteer, shines through the leafless trees in the east after sunset. Accompanying Capella in the south and southeast are the cone-shaped constellation Taurus the Bull and the diminutive Pleiades. Both herald the Yuletide season by evoking images of a Christmas tree and a stray gift wrapped in glittering paper.

By 8 o'clock, you can plainly see Orion in the southeast rousing himself from his annual slumber. His club is held high over his head, his shield brought to bear before the Bull, and his belt stars aligned perpendicular to the horizon. Wait a half hour or so, then follow the belt stars down to the horizon. There you'll see our loyal friend Sirius the Dog Star, its furious flickering alerting us to winter's impending arrival. Procyon in Canis Minor shines to the north, and, now well up in the eastern sky, Castor and Pollux of Gemini the Twins stride onto the celestial stage.

We have come full circle once again. The winter stars are on the rise, and winter's eve is here.

Imparting order in a world of disorder, sublime beauty in an otherwise black night, the stars are there, every year without fail for contemplation and delight. And so they will be for lifetimes to come. In the dim past, the stars were guiding lights. Today they guide us still, inspiring and bolstering lofty dreams, encouraging us to look beyond ourselves, and reminding us of where we came from, and where we shall all, one day, return.

PART II

The Sun, Moon, and Planets

This whole earth which we inhabit is but a point in space. How far apart, think you, dwell the two most distant inhabitants of yonder star, the breadth of whose disk cannot be appreciated by our instruments? Why should I feel lonely? Is not our planet in the Milky Way?

Henry David Thoreau

The skywatcher's Sun

Most everyone knows that the Sun is a star, one that looms large in the sky because it is only 150 million kilometers away. But this school lesson was not always accepted as common knowledge. Ancient civilizations viewed the Sun as distinctly different from the stars, and for obvious reasons. The Sun is large and provides warmth, unlike the stars at night, which are tiny, cold points of light from where we view them. The Sun is also very bright and gives off life sustaining light. This and the fact that it is the most dominant object in the sky made the Sun an object of reverence and worship for many ancient people.

The idea that the stars we see at night are really suns seen across great distances first cropped up in the mid sixteenth century when the Italian philosopher Giordano Bruno suggested that the universe was filled with countless 'suns' and that many of these possibly harbored planets, and, perhaps, life of their own. It was for convictions such as these that Bruno was burned alive at the stake in 1600 in Venice.

Once recognized as a star, the Sun quickly assumed a prominent role in stellar astronomy, becoming the best studied star in the sky. What astronomers have learned about the nature of the myriad other stars stems directly from what they have learned about the Sun.

Though it seems relatively stable and quiescent, the Sun is a dynamic, seething sphere of mostly hydrogen gas well over a million kilometers in diameter. From its innermost to outermost layers, it is a study in thermal contrasts. Its core temperature is 15,000,000 kelvins (27,000,000 °F) but by the time this heat energy reaches the surface, it cools to only 5,800 kelvins (a little over 10,000 °F). Paradoxically, due to magnetic forces that generate heat above the Sun's surface, the temperature in its rarefied outer atmosphere, called the corona, is even hotter than at the surface – at least 2,000,000 kelvins (3,600,000 °F). Altogether, the Sun radiates a staggering 400 million-billion-billion watts of energy per second.

The part of the Sun that we see directly – its visible surface – is called the photosphere. It is upon this stage that the drama of the internal Sun plays out in the form of sunspots and bright filaments called faculae, both of which can be observed with a small telescope equipped with the proper solar filters.

Sunspots and faculae

The most prominent features on the solar disk are sunspots – dark blots that range from Earth-size motes to large complex regions covering billions of square kilometers.

Sunspots are essentially areas where the forces of concentrated magnetic fields inhibit currents of rising hot gas from deep within the Sun. With less hot gas reaching the surface, that part of the Sun is cooler (by about 1,500 kelvins) and thus appears darker. Despite their dark color, sunspots are actually hotter than the surfaces of a great many cooler stars. The spots only appear dark because they are cooler and contrast with their hotter surroundings. If we could somehow place a sunspot in nearby space, it would shine brighter than the Moon.

Sunspots consist of two sections: the penumbra and the umbra. The penumbra forms the outer halo of the sunspot with its feathery brush-stroke appearance. Within the penumbra lies the darker umbra, where the magnetic field forces are strongest. Under high magnification, each sunspot has a character all its own. Some are remarkably symmetrical, resembling the deep yellow petals and black conical center of the North American wild flower, the black-eyed Susan, while others are peanut shaped or completely amorphous. Sunspots can form singly, but most cluster in sunspot 'archipelagoes' with multiple umbrae linked by sinuous dark threads running from one spot to the next. The most complex spots are tell-tale indicators of an intense magnetic field that might spawn a solar flare – a short-lived, intensely bright region caused

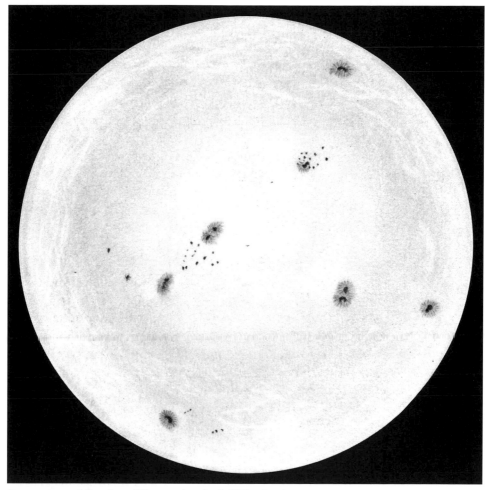

Sunspots and faculae. (Illustration by Jeff Kanipe.)

by the explosive release of energy from within the Sun. Flares are more prevalent during peaks in solar activity and, unfortunately, none except the very brightest are visible without a special narrow-band filter called a hydrogen-alpha filter.

Sunspots develop and decay over a period of two to three weeks, although in some cases, they can last as long as a month or more. In the 12 or so days it takes for sunspots to cross from the eastern limb to the western limb of the Sun, a sunspot group can change shape rapidly. In their development phase, they may swell in size and alter their structure or blossom into complex sunspot groups before dwindling back into the solar granules (the 1,000-kilometer wide convection cells that give the Sun its mottled appearance). If you follow sunspot activity for very long, you'll notice that spots near the Sun's poles take a little longer to transit from limb to limb than those nearer the equator. Though not completely understood by astronomers, this differential rotation is thought to be caused by the Sun's rapidly rotating core.

In addition to sunspots, bright, sinuous filaments called faculae may give the Sun a veined appearance. Faculae (from the diminutive Latin word *facula*, meaning 'little torch') are clouds of tenuous hot gas floating several hundred kilometers above the photosphere. They often precede the formation of sunspots and are commonly observed near a sunspot group. The brightest faculae (only slightly brighter than the photosphere) are harbingers of flares.

Faculae are best seen near the Sun's limb because of a phenomenon called limb darkening. Light from the Sun's limb must pass obliquely through a greater amount of the Sun's atmosphere lying above the photosphere, a region called the chromosphere, to reach the observer. This absorbs photospheric light from the limb. But since faculae actually float above the photosphere, their brightness isn't as suppressed by the chromosphere, making them stand out against the darker limb.

The sunspot cycle

Unlike the energy output of the Sun, the number of sunspots is not constant. Over an average period of 11 years, sunspot numbers increase then decrease, approximately paralleling the increase and decrease in the Sun's overall activity. The fluctuation in sunspot numbers, a phenomenon recognized for over 150 years, is known as the sunspot cycle.

Solar activity peaked in 1989, when sunspots blotched the Sun's disk in greatest numbers. During that time, powerful flares were common, generating aurorae on Earth seen as far south as the Caribbean. One particular flare was so strong it produced a magnetic storm on March 10, 1989, that resulted in power blackouts throughout Quebec and Scandinavia.

Between 1996 and 1997, solar observations indicated a sporadic increase in solar activity and in sunspot numbers, but these reports were considered by astronomers to be more indicative of a prelude to renewed solar activity rather than its onset. Generally, the rise to sunspot maximum is more rapid than its decline. The new solar cycle now underway should translate into a decided increase in sunspot numbers during the first few years of the twenty-first century.

The onset of a new solar cycle can be discerned when the sunspot minimum of a previous cycle is nearing its end. A few new sunspots begin appearing around the lower-middle latitudes of the Sun, about 35° north and south. As the cycle continues, more and more sunspots migrate to lower latitudes until, at sunspot maxima, they form in bands in and around the equator.

Preparing for observing the Sun

Sunspots that cluster into regions at least 50,000 kilometers in diameter can be easily glimpsed with the eye without magnification simply by looking through a protective solar filter. But if you want to see detail in that sunspot, or if you want to see smaller sunspots and faculae, you'll need a telescope. Almost any small telescope is adequate for solar observing, but perhaps the best are quality refractors with 2- to 4-inch apertures, or reflecting telescopes of 4.5 to 6 inches in diameter. Telescopes larger than these tend to resolve roiling air cells, which make the Sun's disk shimmer in and out of focus. Smaller telescopes can easily resolve solar features, but are less effective at resolving atmospheric turbulence.

Without a doubt, the most crucial factor in solar observing is the filter, not only because it allows you to clearly make out features on the Sun, but, more importantly, because it prevents serious eye injury. Here I must present the requisite words of warning. Looking at the Sun with a filter is perfectly safe. But to venture a look, even a quick look, without one is nothing less than reckless. And when you *do* have a filter, be careful. You should think of solar observing as you would rock climbing or scuba diving. It takes concentration, equipment you can trust, and adequate preparation. If you take these precautions, solar observing can be very rewarding, especially during periods when the Sun is active. (More on that later.)

There are two general approaches to observing the Sun: directly, using a solar filter; or indirectly, via projection onto a screen using a small telescope.

One of the safest and least expensive solar filters is made of Mylar, which is available from a number of telescope manufacturers. Mylar is a tough, highly reflective plastic coated with aluminum that, when layered properly, blocks over 99 percent of the Sun's light. You can simply hold the filter over your eyes and look directly at the Sun with no ill effects. (Mylar 'glasses' are available that are made specifically for observing the partial phases of a solar eclipse.) The color of the Sun as rendered in Mylar filters is a light blue, which some people may find distracting, though it doesn't take away from the amount of detail you can see.

Aluminized glass filters are also available from telescope manufacturers. Although more expensive, they have the advantage of being durable and render the Sun in a more realistic yellow-orange color. Their manufacturers also claim that glass filters provide better contrast.

If you use a Mylar or glass filter, make sure it fits snugly over the aperture of your telescope or binoculars. If you use binoculars, you can either fit one filter over one aperture while the other remains capped, or use two filters over both apertures, which is more natural for binocular viewing. Never place the filter between your eye and the eyepiece, where the concentrated light of the Sun will bore a hole right through the filter and into your retina. The object is to block the majority of the Sun's light *before* it enters the telescope so that the light that gets through can be safely observed.

If you don't have a filter, you can project the Sun's image using a common telescope or one side of a pair of binoculars (keep the other aperture capped). This involves allowing the Sun to pass through the telescope unfiltered and projecting the image onto a fairly large, stiff piece of white cardboard placed a couple of feet behind the eyepiece. The projected image can then be easily photographed, and it is a good way for a large group of people to see the Sun at the same time.

There are a couple of cautions with this method, however. First, make sure the Sun doesn't stream through the telescope tube and eyepiece for long periods of time. The intense heat can damage internal tube baffles, crack the glass in your eyepieces, and melt the cement that holds the lens elements together. Every so often, cover the tube opening with an opaque card or dew cap and allow the telescope to cool down for a few minutes. Furthermore, be sure to cap your finder scope to prevent sunlight from inadvertently streaming through and burning your skin or hair, or damaging your eyes.

A note on Schmidt–Cassegrain telescopes. These instruments are especially unsuitable for solar projection because their corrector plates lie close to the focus of the primary mirror. If the light path is not collimated very well, stray sunlight could damage the corrector plate or the tube baffling. Aperture-fitting Mylar or glass filters are the best route here.

When projecting, it is important to shade the screen so you can see the detail without having to squint past the glare reflected back from the white card. If possible, project the image into a darkened tent or shrouded screen. This will give you a comfortable image to observe and one with better contrast. Work at projecting an image that shows a fairly bright disk but with as much detail as possible. For a 3 inch telescope, a projection disk diameter of 6 to 9 inches is best; for 4- to 6-inch telescopes, 10 to 12 inches is fine.

Observing the Sun

An easy way to locate the Sun in your telescope is the shadow method. Insert a low-magnification eyepiece into the eyepiece holder and point your telescope more or less toward the Sun. Then, watching the shadow of the tube on the ground, carefully orient the telescope until its shadow is smallest, meaning the tube will be aligned directly at the Sun and therefore not casting a long shadow. Look in the eyepiece and make more fine-tolerance adjustments to center the Sun's disk. Once the Sun is centered, you can use higher magnification to study the surface in more detail.

Like the Moon, planets, and stars, if the Sun is too low in the sky, you'll be forced to look through excess atmosphere, which can smear out the subtle features and blur even the obvious ones. However, if you wait until the Sun is high in the sky, noon and afterwards, you risk having the Sun's features distorted by the turbulent air cells I mentioned earlier. The best time to look at the Sun through a telescope is during the mid to late morning hours, when it is more than halfway up into the sky, between 45° and 75° in altitude.

Seeing quality at any time can be improved if you do two additional things. First, avoid setting up your telescope on tarmac or concrete. These surfaces absorb and reradiate heat rapidly, which in turn mixes violently with cooler air above the ground, creating turbulence.

Second, keep your telescope shaded. The unrelenting heat of the Sun beating down on the telescope tube generates internal tube currents that can foul the image. You can easily construct a lightweight cardboard collar and mount it near the telescope's aperture so the tube is shaded while observing. Another low-tech alternative is to drape a sheet or tarpaulin across a line strung between two poles that are inserted into the ground or held in place with bricks. The telescope's aperture simply protrudes slightly from a parting or slit in the curtain, while the rest of the instrument is shielded from the Sun. A bit ungainly, perhaps – and not much use under windy conditions or when the Sun is near the zenith – but this method has the advantage of placing the observer in the shade, which not only adds comfort but keeps the eye from being dazzled by the sunlight, something that could prevent you from seeing more subtle detail on the Sun's surface.

Aurorae: the northern and southern lights

Although not categorized with solar phenomena, *per se*, the northern and southern lights, known as the aurora borealis and the aurora australis, respectively, are related to solar activity. As activity increases, sunspots and explosive solar flares appear in greater number. The flares associated with sunspots produce the outflows of charged particles, which in turn create auroral displays as they interact with atoms and molecules in Earth's upper atmosphere.

An aurora usually begins as a dome of reddish or greenish light glowing low over the northern horizon. (In the Southern Hemisphere, the glow first develops in the south.) After an hour or so, the glow may intensify, shaping itself into a distinctive arc that edges southward (northward in the Southern Hemisphere) while shifting across the sky like breeze-blown draperies. During its peak, green, red, and blue rays of light may extend to the zenith and even beyond. The phenomenon may last a few hours or all night.

Throughout recorded history, aurorae have both frightened and inspired people living in the Arctic region. The Inuit of North America believed auroral displays occurred in tandem with daily life. They were the play of unborn children, or the glow cast from torches held by the dead to light the dark of winter for those still living. In Scandinavia, aurorae were interpreted as weather signs. Even today, the language in certain regions there contains terms such as 'weather light' and 'wind light.'

Aurorae are caused by charged particles (protons and electrons) flowing out from the Sun in the solar wind. When the particles reach Earth, they are attracted to the geomagnetic poles, where the magnetic field is concentrated. As the particles accelerate around the magnetic field lines, they excite, or ionize, neutral molecules and atoms present in Earth's atmosphere at an altitude of about 100 kilometers. It is the ion ization of these particles that produces the eerie glow of an aurora.

Because aurorae are produced by the solar wind's interaction with the geomagnetic poles, they can only be seen at high latitudes, typically 60° and higher. During periods of high solar activity, however, aurorae may extend down to latitudes of 40° or less. In 1989, a major solar outburst produced an aurora that was visible in the desert southwest of the United States and as far south as the Caribbean Sea.

The Sun is the brightest, most dominant object visible with the naked eye from Earth. It's presence, or lack of presence, is something everyone notes every day, skywatcher or no. Moreover, we now recognize the Sun as a star seen close up, and we know that stars come in a variety of temperatures, sizes, and types. Had the Sun been only slightly brighter or cooler, had it been a star that underwent extreme fluctuations in energy output, as many stars do, or if it encircled other bright stars, life as we know it could never have arisen on Earth. In that respect, we still have reason to revere the Sun much as our ancestors did. Perhaps more.

Eclipse: greatest shadow show in the solar system

Tables of total solar and lunar eclipses are listed in Appendices I and II, respectively.

An eclipse of the Sun or Moon occurs whenever Earth passes directly between the Sun and the Moon or the Moon passes directly between the Earth and the Sun. In the former case, the Earth casts a broad shadow onto the Moon, creating a lunar eclipse, and in the latter case, the Moon casts its tiny shadow on the Earth, creating a solar eclipse.

Most of the time, orbital geometry prevents these bodies from being in perfect alignment because, rather than lying exactly within the plane of the solar system (i.e., the ecliptic), the plane of the Moon's orbit is inclined 5.1° with respect to the ecliptic. Generally, when the Sun, Moon, and Earth are in a row, the Moon is slightly out of line, being either above or below the ecliptic plane. Thus, the new Moon passes a little above or below the Sun, while the full Moon passes slightly above or below Earth's shadow. This 5.1° inclination prevents us from seeing total lunar and solar eclipses every month.

The Moon's orbital plane and the ecliptic intersect, however, at two crossover points called 'nodes.' A solar eclipse can only occur when the Moon and Sun are at or near the same node; and a lunar eclipse can only occur when the Sun and Moon lie at opposite nodes. The point in the lunar orbit where the Moon moves southward through the ecliptic plane (with respect to Earth's north–south axis) is called the descending node; the point where it moves northward is called the ascending node. Either node will suffice for an eclipse.

Total solar eclipses

A total solar eclipse is one of the most spectacular astronomical phenomena that you will ever witness, this side of a major comet or exploding fireball. The alignment of the Moon in front of the Sun causes the Moon to cast a broad cone-shaped shadow, called the penumbra, onto the Earth. Anyone standing within the penumbra, which is about 6,000 kilometers in diameter, will see a partial eclipse. At the heart of the penumbra lies an even more finely tapered shadow cone, called the umbra. The tip of the umbral shadow cast on to the Earth is usually between 150 and 250 kilometers across near the equatorial regions, and wider as the shadow stretches toward the poles. Anyone standing in the umbra witnesses a total eclipse in which the Sun is completely occulted.

During a total solar eclipse, the Moon's shadow trends west to east across the Earth at about 1,700 kilometers per second; thus the Sun can only be completely eclipsed for a matter of minutes at any one place along the path of totality. Observers first see a small part of the Moon's limb silhouetted against the edge of the Sun's disk at a moment called first contact. The partial phases thus proceed until about an hour to an hour and a half later, when the Moon completely blocks the Sun. This is called second contact, the beginning of totality. The moment just before second contact is often heralded by a dagger-like sunbeam streaming from one spot on the darkened limb. This is known as the diamond ring effect.

As long as the Sun is eclipsed, it is perfectly safe to observe or photograph the darkened disk without solar filters or special protection devices. After all, you're essentially looking at the Moon in silhouette, not the bright surface of the Sun. It is exceedingly dangerous, though, to look at any part of the uneclipsed Sun. A chink of unadulterated sunlight is as bright as any part of the rest of the Sun and can cause permanent eye damage. You should be aware of the duration of the eclipse you're observing, so you can anticipate the Sun's sudden and dazzling reappearance from behind the Moon.

Totality is an uncanny, never-to-be-forgotten experience. The sky grows ever darker until the region immersed in the umbra takes on the hues of twilight. When the Sun is overhead, the sky all around the horizon is tinged with the colors of sunset. Crabs scuttle onto the shore,

A total solar eclipse displaying the diamond ring effect and coronal streamers. (Illustration by Jeff Kanipe.)

birds seek out their nests, roosters crow, and some flowers will even close. If totality occurs along a coastal area, the wind may gust off the water as the temperature suddenly drops. Clouds outside of the path of totality remain brilliantly lit by the partial Sun, but the sky overhead is dark enough to see the planets and the brighter stars.

As totality deepens, the outer atmosphere of the Sun, called the corona, becomes distinctly visible. The corona, comprised of rarefied gases extending millions of kilometers in all directions from the Sun's surface, appears like a milky halo of flower petals. On the limb, you may see bright red tongues of gas arching away from the Sun. These are solar prominences, fountains of energy leaping tens of thousands of kilometers from the Sun's surface.

Third contact occurs some few minutes later (often it seems like only seconds) as sunlight begins streaming through various lunar valleys along the limb, creating another diamond ring or a beautiful string-of-pearls display called Bailey's Beads. The fourth and final contact occurs when the Moon no longer covers the Sun, marking the end of the eclipse.

The fact that we can view such a remarkable event is itself one of Nature's more extraordinary coincidences. Although the Sun is 400 times larger than the Moon, it is also some 400 times further away. As a result, both objects have the same apparent size, half a degree.

Sometimes, however, the Earth is closer to the Sun, as it is in winter months, while the Moon is at or near apogee, its greatest distance from Earth. These circumstances conspire to make the Moon appear slightly smaller than the Sun. When that happens, the Moon can't quite cover the entire solar disk, and the outer limb of the Sun appears as a ring, or annulus, of dazzling light in the sky. This is called an annular eclipse. Although not as dramatic as a total eclipse (and not safe to observe without eye protection) it is still well worth seeing if one occurs in your part of the world. The sky darkens noticeably, and the sight of a ring of sunlight in the sky (viewed through filters or by projecting its image through a telescope) is striking.

Although solar eclipses are predictable events, they don't occur at the same time each year, nor in the same regions of the world. There can be as many as four solar eclipses per year, if you count the partial ones. More often, however, there are two, usually a total and an annular eclipse.

A note on observing a solar eclipse

Observing a solar eclipse can be carried out on many levels. You can use the naked eye, binoculars, a telescope (a small 2- to 3-inch telescope will suffice), or video camera. Whatever you use, though, an absolute must is proper protection for your eyes and instruments. As mentioned above, it is perfectly safe to look upon a fully eclipsed Sun, but never at one that is partially or mostly eclipsed. For details on solar filters for optical systems and the projection method of observing, see 'The skywatcher's Sun' in this book.

The simplest method of observing a solar eclipse is the pinhole camera method. You simply punch a 1-mm hole into a small card and allow the sunlight to stream through the hole onto a second card held 30 centimeters or so away. Sometimes, though, it can be difficult to see the relatively small, dim disk of the Sun in daylight. A variation of this pinhole camera method that alleviates glare is to punch a clean 1-mm hole in one end of a large cardboard box through which the sun's light will enter. On the opposite end, attach a piece of white paper to act as a screen. Holding the box over your head, you stand with your back to the Sun, orienting the hole so the sunlight streams through, and view the projected image on the paper. This method provides both shade and a personal viewing environment, if you don't mind being seen with a large box over your head.

For direct observation and photography, mylar or aluminized glass filters should be used during the partial phases. Mylar 'eclipse glasses,' which you place over your eyes, allow you to look directly at the Sun without fear of injury. Eclipse glasses are available from a number of telescope dealers. You can also use number 14 rectangular arc welder's glass for direct observation. Under no circumstances should you use multiple layers of sunglasses, black and white photographic negatives, or an improper grade of arc welder's glass to look at the Sun.

In the latter stages of partiality, there is another phenomenon worth looking for. Note the mottled appearance of light and shadow under leafy trees. You'll see that the dappling light consists of hundreds of crescent-shaped images of the Sun. The leaves in the tree act like a multi-pinhole camera. A similar effect can be produced by interlacing your fingers and allowing sunlight to pass through the small openings.

Total lunar eclipses

Unlike a solar eclipse, a lunar eclipse is visible over a larger area to everyone who can see the Moon, because its source of illumination – the Sun – is cut off by the Earth.

One of the best ways to view a total lunar eclipse is with binoculars. A powerful telescope will give you a close-up view of the shadow moving slowly over the craters, but with their wide views and sharp optics, binoculars give you a striking perspective on the entire drama of totality.

A lunar eclipse begins when the Moon enters the outer fringe of Earth's shadow, called the penumbra, causing a very subtle diminution of light on the Moon. In the case of penumbral eclipses, the Moon moves only through this outer part of the shadow, which, at the Moon's distance, is over 16,000 kilometers in diameter. Most observers on Earth usually don't notice much difference in the Moon's brightness until a half hour or so into the penumbral phase.

As the Moon penetrates deep into the penumbra, it enters the darker core of Earth's shadow, called the umbra, and that aspect is very noticeable indeed. At this time, the partial phases of the eclipse get underway, as observers watch the curved shape of Earth's shadow gradually close over the Moon. Aristotle, as well as other ancient observers, cited the round shape of Earth's shadow cast on the Moon during a lunar eclipse as one of the earliest proofs that Earth is spherical.

The total phase of the lunar eclipse begins about two and a half hours later as Earth's umbral shadow, which is about 9,000 kilometers across, engulfs the Moon. If the Moon skirts the inside of the cone-shaped umbra, the total eclipse phase will be short, about 45 minutes or less. If it crosses the central part of the umbra, the total phase can last over two hours and the Moon will appear darker.

At totality, the most eerie aspect of a lunar eclipse occurs. The Moon is illuminated by sunlight refracting (bending) through Earth's atmosphere. The blue light is bent the least, but the red light is bent the most, just as it is at sunset or sunrise, and it falls on the Moon. Hence, the Moon takes on a dull, coppery hue. In binoculars, brighter stars can often be seen shining next to the Moon's limb, making the whole scene look like a large ruby set among tiny diamond chips. A lunar eclipse, though not as awe-inspiring as a solar eclipse, nonetheless gives skywatchers an opportunity to see how our nearest neighbor, whose appearance among the stars we take for granted, can be made to look distinctly unusual in its usual place.

The skywatcher's Moon

Except for a few days of the month, the Moon is always visible in the sky somewhere. It has been Earth's constant companion for over 4 billion years and, according to one theory, may have even once been part of our planet until a passsing Mars-size object struck Earth a glancing blow, blasting a piece of it into space. Since then, the airless Moon has taken quite a beating. Its asphalt-colored surface is pitted with countless craters big and small. They and the many prominent faults and craggy mountain ranges make for hours of fascinating scrutiny with a telescope or binoculars.

The Moon lies at a mean distance of 384,000 kilometers and has a diameter of a little over 3,400 kilometers. As moon–planet ratios go in the solar system, our Earth–Moon system is an oddity, with the Moon being rather large for a planet our size. Consequently, a body at the distance and with the mass of the Moon creates some gravitational effects on Earth in the form of flowing tides. The Moon's pull may also have some very minor influence on enhancing earthquakes, although this is not well accepted. Some people claim a full Moon sways behavior. But this is more for the realm of social psychology. Essentially, other than tidal effects and providing light and beauty, the Moon does not overtly affect the Earth.

On the other hand, if we didn't have a Moon, life on Earth would probably not exist, at least not in the form it does today. The Moon provides a gyroscopic stability to the Earth, keeping the poles from tilting dramatically. A world on which the poles flip-flopped would create climatic catastrophes that would wreak havoc for any burgeoning species.

Observing the Moon

The Moon is truly one of the most compelling objects in the night sky. The unaided eye can easily discern a mottled surface of bright and gray patches that arrange themselves into the popular face of the 'Man in the Moon,' or, as seen by the Maya and Aztecs of ancient Mexico, as well as the Mimbres Indians of the southwestern United States, a rabbit. The dark regions are known as maria, Latin for seas, since that is how they appeared to early skywatchers. We know them now as areas where lava spilled across the Moon's surface billions of years ago, probably after an asteroid punched through the thin lunar crust.

As sublime as the Moon appears to the eye, a pair of binoculars turns this marbled world into a wonder. One of the best times to look at the Moon using any optical aid is during its crescent and quarter phases, not when it is full. The reason is simple. When the Moon is full, it is essentially 'noontime' there, with the Sun overhead. Though this geometry makes the Moon appear very bright as seen from Earth, it also lowers the contrast of the lunar features and reduces the shadows cast by mountains, valleys, and craters.

When the Moon is a young or old crescent or at first or last quarter, however, the Sun is rising on the Moon. The shadows and craters cast deep, dark shadows across the surface, and suddenly, the landscape has dramatic relief.

Let's consider the Moon's prominent features at first quarter. Near the upper eastern (right) limb (as the Moon appears to the naked eye and in binoculars), is the round, dark basin known as Mare Crisium. This region can be seen standing in the full glare of the Sun during the early crescent stages. Our perspective of Crisium on the curve of the Moon's limb often gives it an oval appearance. To the west (left) lies Mare Tranquillitatis, the region of the first manned lunar landing. Above and slightly to its left is another round basin, Mare Serenitatis, which stands partly in shadow. Serenitatis is rimmed on its western (or left) edge by great mountain chains.

South of Mare Tranquillitatis, many craters stand out in light and shadow. Most prominent are the trio Theophilus, Cyrillus, and Catharina. Theophilus and Cyrillus are actually overlapping and look like a figure 8. Catharina lies a little below them. Another prominent

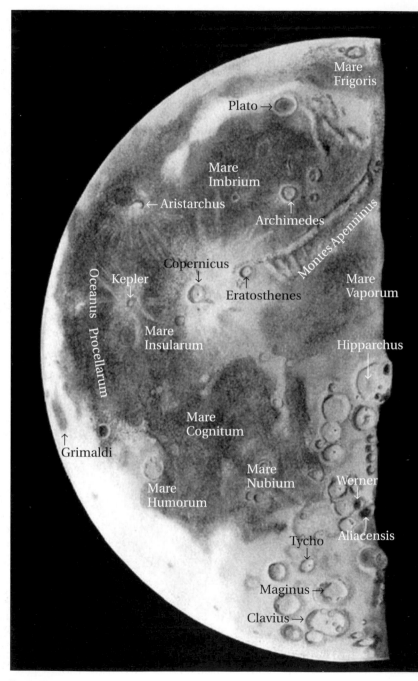

Last quarter Moon.

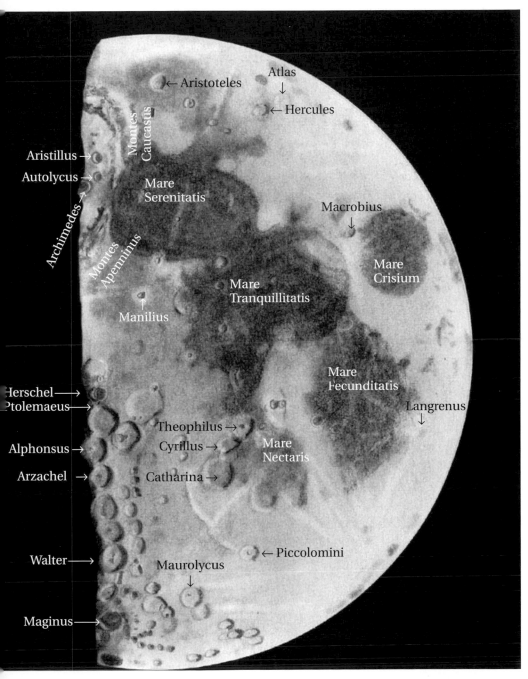

First quarter Moon.

crater, Aristoteles, lies above Mare Serenitatis, near the Moon's north pole. Near first quarter it has a bright west rim, with the floor in deep shadow.

Lined up neatly along the terminator which is the dividing line between the lit and unlit regions of the Moon, just below the lunar equator, are three prominent walled plains. Ptolemaeus lies furthest north (uppermost) and has a number of craters and pits within it. Adjoined to Ptolemaeus to the south, and slightly smaller, is Alphonsus, site of the robotic Ranger 9 landing in March 1965. Finally, just below Alphonsus, is Arzachel, a very conspicuous crater with terraced clefts along its rim and a prominent central peak.

At last quarter, the western (or left) half of the Moon is lit, revealing more dark features. Most prominent among them is Oceanus Procellarum in the upper left quadrant, a sprawling expanse of basalt punctuated by the bright feature Aristarchus, a very recent impact crater. Observers with telescopes have reported unusual brightness and color changes in this feature, some of which may be attributable to gas emissions from within the crust.

Further east (right) is Mare Imbrium, a large, round basin that makes up one of the 'eyes' of the Man in the Moon. Two bright ray craters, Copernicus and Kepler, lie near the equator in this section. The rays are produced by material ejected when an asteroid or comet struck the surface, splashing out bright fresh material from below. Mare Nubium lies directly below, or south, of Imbrium in the Moon's southern hemisphere. Mare Humorum, one of the smallest of the maria, lies to the west, or left.

Most of the Moon's southern hemisphere is rugged and pockmarked with craters, including the bright ray crater Tycho. To its south lies Clavius, one of the most dominant walled plains in this sector of the Moon. Its rim and floor are pitted with a variety of craters.

Predicting moonrise

Many people get confused when it comes to predicting when and where the Moon will appear in the sky. Unlike the Sun, the Moon doesn't rise and set daily, nor does it appear in the same part of the sky on a nightly basis. Moreover, sometimes it appears in the the evening, while at other times it's in the morning. The Moon just seems all over the sky. Once you learn a few lunar basics, though, it is not difficult to master the comings and goings of our nearest neighbor in space.

First, the Moon moves eastward in the sky, despite the fact that we see it carried along with the stars toward the west. The overt westward motion, like the rising and setting of the stars, is due to Earth's rotation. The Moon's more subtle eastward movement, however, is a result of its orbit around Earth. This eastward motion can be seen if you carefully note the Moon's position with respect to bright background stars over the course of a few nights. This task is easier when you observe the Moon between crescent and first-quarter phase. If you observe the Moon at the same time every night, you'll see that it has a daily motion of about 15° eastward – that's about one and a half times the width of your fist held at arm's length.

Second, since Earth rotates toward the east and the Moon orbits Earth toward the east, it takes a bit longer each day for any location on Earth to rotate around so that the Moon has the same altitude above the horizon that it had the previous night. On average, this delay is about 50 minutes; thus the Moon rises later each day by roughly 50 minutes. Bear in mind, however, that this period is abbreviated in the autumn at high northerly latitudes. (See the section on the Harvest Moon for more on this exception.)

Finally, the Moon displays a phase because it shines by reflected sunlight. As the Moon orbits Earth, the lighting angle changes in predictable ways, moving from crescent to quarter to full and back again. If you know the lunar phase, you can predict with some accuracy when the Moon will rise and set.

For example, when the Moon is at quarter phase (or half lit), it must be 90° from the Sun in our sky. In other words, the Moon is either on the meridian as the Sun sets, in which case the Moon sets about midnight, or is on the meridian as the Sun rises, in which case it sets around noon. You'll never see a half-lit Moon in the west just after sunset or setting in the west just before sunrise. It's geometrically impossible.

When the Moon is full, it must always appear 180° opposite the Sun in the sky. In other words, a full Moon rises at sunset, is on the meridian at midnight, and sets when the Sun rises in the morning. You'll never see a full Moon on the meridian at sunset. If, however, you see a half-lit Moon rising in the east at sunset, then you must assume that it is undergoing a lunar eclipse.

Knowing that the Moon rises about 50 minutes later each night allows you to make a rough prediction for moonrise simply by knowing which phase the Moon is in. You need only one additional fact: the period from new Moon to new Moon (called a synodic month) is 29.5 days, so the interval of days between the four main phases of the Moon is approximately 7 days. If the Moon is 3 days past full – when it rose at sunset – then it will rise about 150 minutes after sunset (3×50), or about two and a half hours later. That also means that the Moon will set about this much time after sunrise.

Once you can relate lunar phases to lunar position, the Moon's schedule becomes almost second nature, making moonrise a more predictable monthly event.

The tides

Fishermen, harbormasters, and coastal dwellers around the world can't help but notice an astronomical phenomenon that affects their region, sometimes dramatically, every day: the ebb and flow of the tides. The Earth's daily rotation, along with the Moon's incremental eastward revolution, produces two high tides and two low tides during each 24-hour period. Because the Moon rises approximately 50 minutes later each day, high tides occur about 50 minutes later on successive days, at intervals of 12 hours and 25 minutes.

Tides are governed by the Moon's gravitational effects on the Earth and its oceans. The Sun, too, has an effect, as does the centrifugal force of Earth's revolution, but it is the Moon that exerts the most influence. Though the Sun is nearly 400 times larger than the Moon, it is also about 400 times further away, which is why both bodies appear to have the same diameter in the sky. Since gravitational attraction varies inversely with the square of the distance between two bodies, the Moon exerts 46 times more gravitational pull than the Sun.

Contrary to what many people think, the Moon does not 'pull' water away from the Earth. Rather, high tides result when the Moon's gravity causes the ocean to flow over the Earth's surface toward the region where the Moon appears directly overhead. However, because of the friction between the ocean and the ocean floor, the impeding effects of land masses, differences in water depth, wind, and Earth's rotation, the water 'heaps up,' not directly below the Moon, but somewhat behind the Moon's zenithal position. Taken together, these circumstances conspire to create oscillations in the ocean surfaces, so that the water over a large area rises and lowers in cycles.

Additionally, as high tide occurs on the Moon side of the Earth, a corresponding high tide takes place on the opposite side of the Earth. This apparent paradox is explained by the fact that the Earth's surface on the lunar side of our planet is much closer to the Moon than its center. Consequently, the Moon's gravity attracts the nearer surface more strongly than its center. At the same time, however, the Moon attracts the center more strongly than the ocean on Earth's far side. Thus, the Moon's gravity pulls the rigid Earth away from the less-rigid ocean on the opposite side, causing a high tide there. It's akin to what happens when you pull a plastic bowl of water toward yourself a little too quickly. The bowl itself flexes slightly from the force of the pull, but the water in the bowl, being less resilient, is more easily displaced in the opposite direction.

On a daily basis, one tidal cycle is very much like another. From its lowest point, the water level rises gradually for about six hours, then lowers again for another six hours until it reaches low tide. When the Moon is new or full, and nearest to the Earth for the month (perigee), exceptionally high tides, called spring tides, occur. They are most noticeable in funnel-shaped bays and estuaries. In places like the Bay of Fundy, between Nova Scotia and New Brunswick, Canada, the tides can rise by as much as 15 meters (50 feet).

When the Moon is at right angles to Earth and the Sun, as when the Moon is at first or last quarter, and is also at its greatest distance from Earth for the month (apogee), the attraction is not as strong and exceptionally low tides, called neap tides, occur.

Exceptionally high spring tides occur when the new or full Moon is at perigee and Earth is also nearest the Sun (perihelion). Under those circumstances, both the Moon and Sun act together to increase the tide levels. With the relevant information, you can predict you own 'super spring tide' event. Times of high and low tide usually can be found in your local newspaper in the weather section. The dates of lunar perigee and apogee for the year, as well as the dates of the phases, can be found in any current almanac. The date of perihelion falls at nearly the same time every year, usually around January 2 or 3. (Aphelion, Earth's greatest distance from the Sun, is around July 4 or 5.)

The more coincident least-lunar perigee, perihelion, and the new Moon are, the greater the height of the spring tide. For example, in the year 2005, the Moon reaches its least perigee distance on January 10, when it will also be at the new phase. Moreover, the Earth–Moon system will be nearest the Sun that year on January 2. With all these events happening within nearly the same time period, the spring tides of January 2005 should be exceptionally high.

The Harvest Moon

Autumnal moonlight was a highly valued commodity before the invention of electric lights, particularly in the northern parts of the world when crops had to be gathered from the fields before the first frosts. The Moon's extra light allowed workers to harvest into the evening hours, long after the Sun had set. In fact, this Moon has come to be known as the Harvest Moon.

By loose definition, the Harvest Moon is the full Moon nearest the autumnal equinox. Most years, the title goes to the full Moon in September, since the equinox occurs around September 22. Sometimes, however, the interval of days between the full Moon in September and the autumnal equinox is greater than that of the equinox and the full Moon in October (as was the case in 1998). In such a case, then, the October full Moon is decreed the Harvest Moon.

Some people insist on calling the October Moon the Harvest Moon, no matter where it falls in relation to the equinox. The October Moon, however, is more properly known as the Hunter's Moon because, according to lore, its extra light helps hunters track animals at night among the husks and chaff of harvested fields.

At any rate, you may well ask how the Moon can provide 'extra' light around the equinox. The answer lies both in the clockwork motions of the solar system and the Earth's 23.3° tilt.

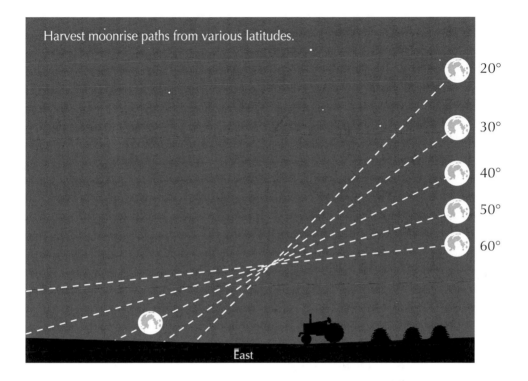

Harvest moonrise paths from various latitudes.

In autumn, the setting Sun is at the autumnal equinox, a point in the sky that lies halfway between its northernmost and southernmost positions for the year. In the meantime, on the opposite side of the sky the rising full Moon is at the vernal equinox, or where the Sun will be on the first day of spring. As it so happens during this time of year, the ecliptic – the apparent annual path across the sky followed by the Sun and Moon – is more inclined toward the horizon the further north you go. At latitude 30°, for example, the ecliptic makes a 45° angle with the eastern horizon, but at latitude 50° that angle shrinks to 20°.

The shrinking ecliptic angle means the Moon's eastward motion does not take it as far below the eastern horizon from day to day as it does at other times of the year. The effect is particularly noticeable at high northerly latitudes. At latitude 30° N, the daily difference in moonrise times around Harvest Moon is about 40 minutes. At 50°, the difference dwindles to about 30 minutes, and at 60° the difference amounts to only 20 minutes.

Thus, there is less difference in moonrise times from night to night for about a week at latitudes above 40° N around the autumnal equinox. This means that the Moon is up in the sky at nearly the same time on successive evenings and is in the sky all night, providing extra light.

If a shallow ecliptic angle can create a Harvest Moon phenomenon in the Northern Hemisphere, a similar Harvest Moon effect must also occur in the Southern Hemisphere for the same reason. Indeed it does, but it occurs in March (i.e., southern autumn), when the Moon is at the autumnal rather than the vernal equinox, and that part of the ecliptic is most inclined toward the eastern horizon.

As in the Northern Hemisphere, the effect is more pronounced the further south you go. In places like Sydney and Melbourne (33° S and 38° S, respectively) successive full Moon rises differ by about 30 to 35 minutes. At latitude 42° S, the difference amounts to about 25 minutes. To really notice the Moon rising each night at nearly the same time, however, you'd have to be on a ship at sea or standing on Antarctica. At latitude 60° S, for example, the difference in consecutive moonrises is only 20 minutes.

The Harvest Moon is not strictly an American tradition. In Chinese communities in the United States, the Harvest Moon festival is a significant holiday that, apparently, can even influence death rates. Carl Sagan, in his book *The Demon Haunted World*, cites one study that showed that deaths in the Chinese community fell by 35 percent during the week preceding the festival but jumped by that amount the following week. A close examination of the data found that the fluctuation in the death rate occurred mostly among women 75 years and older. Traditionally, the Chinese Harvest Moon festival is presided over by the oldest women of the house. Apparently, these women managed to ward off death in some psychosomatic way for a couple of weeks in order to fulfill their ceremonial duties. Such is the control of the Moon in our lives.

Blue Moons

Most people are familiar with the expression 'once in a blue Moon' and take it for granted that it means 'once in a great while' or 'rarely.' It is one of those reflexive, oft-spoken statements that convey meaning, though, when examined by itself, has little or no intuitive meaning.

Actually, there are two definitions to this enigmatic term, and either or both may be cited in popular astronomical literature. Professional astronomers, when they deign to mention blue Moons in texts, refer to the occasional blue color of the Moon resulting from the scattering effects of high-altitude dust lofted by forest fires or volcanic eruptions. Indeed, a blue Moon was widely observed September 26, 1950, the result of an extensive Canadian forest fire, and also August 27, 1991, from ash elevated by the volcanic eruption of Mount Pinatubo in the Philippines.

Recreational astronomers and skywatchers put a different spin on the blue Moon definition. Most of the time, each month has only one full Moon phase. Occasionally – one might even say rarely – a month will have two full Moons: one at the very beginning and one at the very end. Accordingly, a blue Moon is 'the second full Moon in a month.' Such was the case in August 1993. The first full Moon occurred on August 2; the second on August 31.

Of course, there's a slight agitation to the dual full Moon definition of blue Moons that muddies the waters a bit. Clearly, the month in which we experience a blue Moon depends on where we are located on Earth or, more specifically, where we are at the time the Moon's full phase officially occurs.

For example, in the August 1993 case, the Moon became full the first time on August 2 at 12 hours 10 minutes Universal Time, which is the mean solar time measured at the prime meridian in Greenwich, England. In the United States, however, when daylight saving time is employed, there is a four-hour time difference between the East Coast time zone and Greenwich, England (five hours when daylight saving time is not in effect). So, on the east coast, the Moon reached its full phase at 8:10 a.m., Eastern Daylight Time, August 2, 1993.

Here's the tricky part. The Moon did not reach its full phase again until 2 hours 33 minutes Universal Time, September 1. But don't forget about that 4-hour time difference between Greenwich, England, and the eastern U.S. When you factor that in, we essentially step 'backward' (westward) in time to 10:33 p.m., August 31. So, those living in North America got a blue Moon – with an hour and a half to spare – while those living in Europe did not (although they did get one in September of that year).

On average, a full Moon occurs every 29.53 days. Divide the number of days in a tropical year – 365.2422 – by 29.53 and you get the number of full Moons in a year: 12.369. The extra amount has to crop up somewhere, and it does, on average, every 2 years plus 7, 8, 9, or 10 months, when a month contains two full Moons.

The following list features upcoming years with months having two full Moon phases, thereby constituting a blue Moon. The world regions noted fall largely within the time of the full Moon phase for each date.

2001
October 1 and October 31 – North and South America
October 2 and October 31 – Europe
November 1 and November 30 – Australia/Asia, Africa

2004
July 1 and July 31 – North and South America
July 2 and July 31 – Europe, Africa, Australia/Asia

2007
May 1 and May 31 – Europe, North and South America
May 2 and May 31 – Africa
June 1 and June 30 – Australia/Asia

2009
December 1 and December 31 – Europe, North and South America, Africa
December 2 and December 31 – Australia/Asia

2012
August 1 and August 31 – Europe, North and South America, Africa
August 2 and August 31 – Australia/Asia

2015
July 1 and July 30 – North and South America
July 2 and July 31 – Europe, Asia/Australia

2018
January 1 and January 31 – North and South America, Africa, Europe
January 2 and January 31 – Australia/Asia
March 1 and March 30 – North and South America
March 1 and March 31 – Africa, Europe, Australia/Asia

2020
October 1 and October 30 – Europe, North and South America, Africa
October 1 and October 31 – Australia/Asia

2023
August 1 and August 30 – North and South America, Europe, Africa
August 1 and August 30–31 – Australia/Asia

2026
May 1 and May 30 – Europe, North and South America, Africa
May 1 and May 31 – Australia/Asia

Sometimes, the contrivance of our calendar prevents a two full Moon-month. For example, the year 2048 sees a double full Moon in January. It would also have one in March, except that 2048 is a leap year, so the full Moon falls on February 29, not March 1, with the subsequent full Moon phase occurring March 30.

The year 2018 is unusual in having two months of dual full Moons. A similar situation last occurred in 1999, when the Moon was full twice in January (in Europe) and again in March (in Europe and western North America). After 2018, we won't see a double-double of full Moons until January and March of 2037 (for both Europe and North America). I suppose that would make these 'blue, blue' Moons the rarest of all.

Finding and observing the planets

Much has been written in this book on how to find bright stars and constellations, observe meteor showers, and other cyclical sky phenomena. But almost no mention has been made about the objects that garner some of the greatest interest among skywatchers: the planets. That's not to say that the planets aren't always in the sky, because they are. However, they aren't always in the same part of the sky year after year. That's the tricky part.

The planets were known as 'wanderers' to ancient astronomers because that's exactly what they appear to do: wander back and forth among the constellations of the ecliptic, or zodiac. No naked-eye planet stays in one region of the ecliptic forever because the planets, like Earth, are in orbit around the Sun. As each planet traces out its varied orbit, it shifts nightly with respect to the distant background stars. Planetary orbits are well established, however, allowing their future positions to be easily plotted. You can obtain these positions from the various sky almanacs listed in the bibliography, or from astronomy software programs like Voyager, The Sky, and Starry Night. (These and other ephemeris programs are available from most companies offering books and software for amateur astronomers. Some primary vendors are Sky Publishing Corporation, P.O. Box 9111, Belmont, Massachusetts 02178-9111, (www.skypub.com/); the Astronomical Society of the Pacific, 390 Ashton Avenue, San Francisco, California, 94112 (www.aspsky.org); and Orion Telescopes and Binoculars, P.O. Box 1815, Santa Cruz, California 95061 (www.oriontel.com). They may also be available from nature and science stores found in shopping malls around the world.)

In lieu of these sources, I've provided tables in Appendix III to help you in tracking, in a general sense, planet positions until the year 2010. The tables are not, however, meant to be taken as a comprehensive guide, because to provide a decade's worth of daily planetary positions would take many pages, which is well beyond the scope of this book.

The outer planets are easier to track because they can linger in one region of the sky for months; Uranus, Neptune, and Pluto, for years. Not so the inner planets. Mercury lives up to its name not only by being difficult to see (it never strays further than 28° from the Sun), but by never remaining in the same part of the sky for more than five or ten days. On January 1, 2008, for example, Mercury is in the evening sky, very low in the west right after sunset in Sagittarius. By April 1, however, it is low in the eastern morning sky in Pisces. In a month's time, Mercury can bound through as many as three or four constellations.

Sometimes a planet lies too near the Sun in the sky to be seen at all. Mercury and Venus are often blocked from view by being in the Sun's glare on the far side, an aspect known as superior conjunction, or on the near side of the Sun, in inferior conjunction. The outer planets, obviously, can only be hidden by lying too near the Sun when they pass into superior conjunction.

Observers have their work cut out for them in keeping up with Mercury's whereabouts. In the Northern Hemisphere, the best time to look for it in the evening sky is in the spring, when the ecliptic is most inclined to the horizon. Autumn is the most favorable time to look for Mercury in the morning sky, again because of the ecliptic's steep inclination at that time of year. In the Southern Hemisphere, the optimal time to look for Mercury in the evening sky is during the southern spring, and in the morning sky, during the southern autumn.

Of all the planets, Venus is the most obvious. Being the third-brightest object in the sky, after the Sun and the Moon, it is often brilliant enough to cast faint shadows onto light-colored backgrounds. The clouds enshrouding the planet are very effectual at reflecting light, which is why the planet is sometimes mistaken for the blazing landing beacon of an incoming passenger jet. When not in superior or inferior conjunction, Venus can be seen either in the west in twilight as an 'evening star' or in the east just before dawn, as a 'morning star.'

On rare occasions Venus can appear as both an evening and morning star. About every eight years – or every five inferior conjunctions – the planet takes a higher-than-usual path north of the Sun, allowing it to be seen through strong twilight just east of the Sun at sunset and west of the Sun before sunrise on the same night. The next evening-morning apparation of Venus occurs in 2001; thereafter, 2009.

Venus is dazzlingly bright in a small telescope but, owing to its cloud cover, virtually featureless. Both Venus and Mercury, however, display distinct phases, not unlike the Moon's crescent, gibbous, and quarter phases. When the planets are at their greatest angular distance from the Sun, or elongation, they appear in the gibbous and quarter phases. As they edge closer toward the Sun, their phases shrink to crescents. Venus' phase changes are the more dramatic of the two as it approaches inferior conjunction. As the phase dwindles, its overall disk swells enormously, because the planet is swinging toward Earth. At quarter phase during greatest elongation, Venus may be 25 arcseconds in apparent diameter, but just before inferior conjunction, it can display a disk over twice that size.

Unlike Mercury and Venus, which can only be seen just west or east of the Sun in the morning or evening sky, the outer planets – Mars, Jupiter, Saturn, Uranus, Neptune, and Pluto – can be seen anywhere along the ecliptic any time of night. Being the brighter bodies, Mars, Jupiter, and Saturn can easily be spotted as an extra 'star' in an ecliptic constellation. And because their motions are not as great per month as the inner planets, their positions are easy to track from month to month once you've identified them.

The outer planets are easiest to locate and observe when they come into 'opposition.' This aspect occurs when the planet lies opposite the Sun in the sky (much as the Moon does when it is full, rising as the Sun sets). During this time, the planet is fully illuminated by the Sun and is also nearer to Earth than it is during the rest of the year. Thus it is at its brightest. Opposition is an excellent time to observe a planet in a telescope because its disk is largest then, and the most detail can be discerned. Opposition dates for the outer planets are provided in Appendix IV.

Mars, around the time of its opposition, is a compelling telescopic object. Its deep ocher color is punctuated with dark surface features that extend across the disk. The most prominent feature is wedge-shaped

Syrtis Major, which can be seen along the planet's equatorial zone. To its south is the bright circular Hellas basin, which is essentially filled with bright sand. Adjacent to Syrtis Major is the Meridiani Sinus, an oval dark plane. And, depending on which hemisphere is inclined toward Earth, the planet's polar cap can easily be discerned as a bright spot near the planet's limb.

Blowing sand and global dust storms often cause these mapped features to change shape, brighten, or darken. Sometimes, you can actually see dust storms develop. They start out as small bright regions that, over the course of several days, may spread across the entire planet obscuring all surface detail.

Mars' two moons, Phobos and Deimos, unfortunately cannot be seen in a small telescope and are difficult in even large instruments.

Jupiter and Saturn are not terrestrial planets, like Mars and Earth, but giant gaseous spheres. In a telescope, you can easily see cloud belts and bands girdling both planets, but particularly Jupiter. The belted structure of that planet arises from jets of wind circulating in alternating directions around the planet.

Jupiter might very well be considered its own solar system. First, because of its great internal pressures, the planet radiates more heat than it receives from the Sun. In fact, if Jupiter were 90 times more massive, it would be classified as an extremely low-mass star. Second, it has its own complement of 'planets' in the form of its 16 moons. The four largest – Io, Callisto, Ganymede, and Europa – were first seen by Galileo in his telescope. Collectively, they are still referred to as the Galilean satellites. In a backyard telescope, these can be seen as small stars lined up on either side of the planet. Over the course of an hour, you can see the moons gradually shift positions. Sometimes, one or more may not be seen because they are behind the globe. But when one of the moons passes between the Sun and the planet, it casts a shadow, seen as a dark dot, on the planet's cloud tops. These shadow transits can be observed in telescopes as small as 2 and 3 inches in diameter.

The most intriguing feature of Saturn, of course, is its magnificent ring system, which any small telescope shows clearly. At moderate magnifications, you can see the rings passing behind the globe with the planet's shadow cast dramatically across the ring plane. Look closely and you may be able to see a dark, narrow gap in the rings, known as

Cassini's division (named for Giovanni Cassini, the Italian astronomer who noted the feature in 1675). The rings consist of billions of boulder- to dust-size particles of water ice and rock. Over the course of several years, the planet's 27° inclination tilts one face of the rings and then the other toward Earth. Every 15 years, Earth passes directly through the planet's ring plane and the rings nearly vanish, except for a needle-thin line that appears to impale the globe.

Saturn, too, has a number of satellites – at least 18. The largest, Titan, has an atmosphere and, as such, reflects more sunlight. Consequently it is Saturn's brightest moon. It appears as a tiny star a few 'ring diameters' off to one side of the planet. Four other moons – Rhea, Dione, Tethys, and Enceladus – can sometimes be discerned in 4- to 8-inch telescopes, hovering near the planet like moths around a distant lamp light.

Uranus, Neptune, and Pluto, while they may inhabit their host con- stellation for the longest periods of time, are the most difficult of the planets to observe, much less find. Nevertheless, I've included their locations in Appendix III for those with access to a telescope or who just want to know approximately where in the sky these planets are.

Uranus, which is just visible to the naked eye in dark skies (magni- tude 5.6) is most easily found using binoculars. It has a small but dis- tinctly greenish disk. Small telescopes show a slightly larger, but featureless, disk. Seeing Neptune (magnitude 7.8) requires a telescope and a detailed finder chart, which is available in most popular astro- nomical almanacs and astronomy publications. You can distinguish Neptune as a bluish dot among the stars.

Pluto requires at least an 8-inch telescope, medium magnification, and a very detailed finder chart. At nearly magnitude 14, it is little more than the faintest star you can see in a field of stars. You can identify Pluto by noting the positions of the stars in the telescope's field of view over several nights. The star that shifts its position is Pluto.

Appendices

Appendices

Appendix I

Total and annular solar eclipses: 2001 to 2024

Below is a list of total and annular eclipses visible until the year 2024. Note: the duration listed is how long the Sun is eclipsed by the Moon at the given longitude and latitude of greatest eclipse.

2001   June 21: Total solar eclipse
Site of greatest eclipse: Lat. 11° 15.8S; Long. 2° 44.6E
Duration at site of greatest eclipse: 4 minutes, 56 seconds
Path of totality: South Atlantic to southern Africa

December 14: Annular solar eclipse
Site of greatest eclipse: Lat. 0° 37.3N; Long. 130° 41.4W
Duration at site of greatest eclipse: 3 minutes, 53 seconds
Path of annularity: mid Pacific Ocean; Central America

2002   June 10: Annular solar eclipse
Site of greatest eclipse: Lat. 34° 32.7N; Long. 178° 36.7W
Duration at site of greatest eclipse: 22 seconds
Path of annularity: central North Pacific Ocean

December 4: Total solar eclipse
Site of greatest eclipse: Lat. 39° 28.2S; Long. 59° 33.6E
Duration at site of greatest eclipse: 2 minutes, 3 seconds
Path of totality: southern Africa; Indian Ocean; western Australia

2003   May 31: Annular solar eclipse
         Site of greatest eclipse: Lat. 66° 49.4N; Long. 24° 0.2W
         Duration at site of greatest eclipse: 3 minutes, 36 seconds
         Path of annularity: southern Greenland; Iceland

         November 23: Total solar eclipse
         Site of greatest eclipse: Lat. 72° 40.6S; Long. 88° 29.4E
         Duration at site of greatest eclipse: 1 minute, 57 seconds
         Path of totality: eastern Antarctica

2005   April 8: Annular/Total solar eclipse
         Site of greatest eclipse: Lat. 10° 34.6S; Long. 118° 57.7W
         Duration at site of greatest eclipse: 42 seconds
         Path of annularity/totality: South Pacific Ocean; Panama; northern
         South America

         October 3: Annular solar eclipse
         Site of greatest eclipse: Lat. 12° 52.4N; Long. 28° 45.0E
         Duration at site of greatest eclipse: 4 minutes, 31 seconds
         Path of annularity: northwestern to eastern Africa; Indian Ocean

2006   March 29: Total solar eclipse
         Site of greatest eclipse: Lat. 23° 8.6N; Long. 16° 45.9E
         Duration at site of greatest eclipse: 4 minutes, 6 seconds
         Path of totality: South Atlantic Ocean; north-central Africa; eastern
         Mediterranean; Turkey; southern Russia

         September 22: Annular solar eclipse
         Site of greatest eclipse: Lat. 20° 39.9S; Long. 9° 3.4W
         Duration at site of greatest eclipse: 7 minutes, 9 seconds
         Path of annularity: southern Atlantic Ocean

2008   February 7: Annular solar eclipse
         Site of greatest eclipse: Lat. 67° 34.6S; Long. 150° 27.0W
         Duration at site of greatest eclipse: 2 minutes, 11 seconds
         Path of annularity: extreme southern Pacific Ocean; western Antarctica

August 1: Total solar eclipse
Site of greatest eclipse: Lat. 65° 38.2N; Long. 72° 16.3E
Duration at site of greatest eclipse: 2 minutes, 27 seconds
Path of totality: northern Greenland; Arctic Ocean; northern Siberia

2009   January 26: Annular solar eclipse
Site of greatest eclipse: Lat. 34° 4.7S; Long. 70° 16.4E
Duration at site of greatest eclipse: 7 minutes, 53 seconds
Path of annularity: southern Indian Ocean; western Indonesia

July 22: Total solar eclipse
Site of greatest eclipse: Lat. 24° 12.0N; Long. 144° 8.4E
Duration at site of greatest eclipse: 6 minutes, 39 seconds
Path of totality: eastern China; extreme southern Japan; northern Micronesia; western North Pacific Ocean

2010   January 15: Annular solar eclipse
Site of greatest eclipse: Lat. 1° 37.3N; Long. 69° 20.2E
Duration at site of greatest eclipse: 11 minutes, 0 seconds
Path of annularity: eastern Africa; Indian Ocean; Thailand

July 11: Total solar eclipse
Site of greatest eclipse: Lat. 19° 46.5S; Long. 121° 51.0W
Duration at site of greatest eclipse: 5 minutes, 20 seconds
Path of totality: French Polynesia; South Pacific Ocean

2012   May 20: Annular solar eclipse
Site of greatest eclipse: Lat. 49° 4.6N; Long. 176° 19.1E
Duration at site of greatest eclipse: 5 minutes, 46 seconds
Path of annularity: East China Sea; southern Japan; south of Aleutian Islands; northwest coast United States

November 13: Total solar eclipse
Site of greatest eclipse: Lat. 39° 457.6S; Long. 161° 17.9W
Duration at site of greatest eclipse: 4 minutes, 2 seconds
Path of totality: extreme northern Australia; northeast of New Zealand; South Pacific Ocean

2013   May 10: Annular solar eclipse
Site of greatest eclipse: Lat. 2° 12.3N; Long. 175° 30.5E
Duration at site of greatest eclipse: 6 minutes, 3 seconds
Path of annularity: northern Australia; southern New Guinea; Solomon
Islands; southern North Pacific Ocean

November 3: Annular/Total eclipse
Site of greatest eclipse: Lat. 3° 29.4N; Long. 11° 39.6W
Duration at site of greatest eclipse: 1 minute, 39 seconds
Path of annularity/totality: South Atlantic Ocean; southern central-
Africa

2015   March 20: Total solar eclipse
Site of greatest eclipse: Lat. 64° 24.9N; Long. 6° 34.0W
Duration at site of greatest eclipse: 2 minutes, 46 seconds
Path of totality: Norwegian Sea; Svalbard, Norway; Arctic Ocean

2016   March 9: Total solar eclipse
Site of greatest eclipse: Lat. 10° 6.5N; Long. 148° 50.2E
Duration at site of greatest eclipse: 4 minutes, 9 seconds
Path of totality: eastern Indian Ocean; Indonesia; South Pacific Ocean

September 1: Annular solar eclipse
Site of greatest eclipse: Lat. 10° 41.4S; Long. 37° 48.1E
Duration at site of greatest eclipse: 3 minutes, 5 seconds
Path of annularity: southern Africa; Madagascar; western Indian Ocean

2017   February 26: Annular solar eclipse
Site of greatest eclipse: Lat. 34° 41.6S; Long. 31° 8.4W
Duration at site of greatest eclipse: 44 seconds
Path of annularity: southern Chile and Argentina; mid South Atlantic
Ocean; Angola

August 21: Total solar eclipse
Site of greatest eclipse: Lat. 36° 58.0N; Long. 87° 37.7W
Duration at site of greatest eclipse: 2 minutes, 40 seconds
Path of totality: North Pacific Ocean; northwestern to southeastern
United States; Atlantic Ocean

2019   July 2: Total solar eclipse
        Site of greatest eclipse: Lat. 17° 23.5S; Long. 108° 57.0W
        Duration at site of greatest eclipse: 4 minutes, 32 seconds
        Path of totality: southern South Pacific Ocean; Chile; Argentina

        December 26: Annular solar eclipse
        Site of greatest eclipse: Lat. 0° 59.7N; Long. 102° 18.4E
        Duration at site of greatest eclipse: 3 minutes, 39 seconds
        Path of annularity: mid Arabian Peninsula; southern India; Indonesia;
        Micronesia

2020   June 21: Annular solar eclipse
        Site of greatest eclipse: Lat. 30° 31.0N; Long. 79° 43.2E
        Duration at site of greatest eclipse: 38 seconds
        Path of annularity: Arabian Peninsula; northern India; Nepal; southern
        China

        December 14: Total solar eclipse
        Site of greatest eclipse: Lat. 40° 21.1S; Long. 67° 54.4W
        Duration at site of greatest eclipse: 2 minutes, 9 seconds
        Path of totality: South Pacific Ocean; Chile; Argentina; South Atlantic
        Ocean

2021   June 10: Annular solar eclipse
        Site of greatest eclipse: Lat. 80° 48.9N; Long. 66° 36.4W
        Duration at site of greatest eclipse: 3 minutes, 51 seconds
        Path of annularity: Siberia; Arctic Ocean; extreme northern Canada;
        Baffin Bay; western Greenland; Hudson Bay

        December 4: Total solar eclipse
        Site of greatest eclipse: Lat. 76° 45.8S; Long. 46° 17.6W
        Duration at site of greatest eclipse: 1 minute, 54 seconds
        Path of totality: Weddell Sea; West Antarctica

2023   April 20: Annular/Total solar eclipse
        Site of greatest eclipse: Lat. 9° 36.0S; Long. 125° 50.5E
        Duration at site of greatest eclipse: 1 minute, 16 seconds
        Path of annularity/totality: southern Indian Ocean; extreme west coast

of Australia; Lesser Sunda Islands; western New Guinea; southern Micronesia; South Pacific Ocean

October 14: Annular solar eclipse
Site of greatest eclipse: Lat. 11° 21.2N; Long. 83° 2.8W
Duration at site of greatest eclipse: 5 minutes, 17 seconds
Path of annularity: northwestern to south central United States, Yucatan; Central America; northern South America

2024   April 8: Total solar eclipse
Site of greatest eclipse: Lat. 25° 16.9N; Long. 104° 5.1W
Duration at site of greatest eclipse: 4 minutes, 28 seconds
Path of totality: South Pacific Ocean; southwestern Mexico, south central to northeastern United States; Newfoundland; North Atlantic Ocean

October 2: Annular solar eclipse
Site of greatest eclipse: Lat. 21° 58.1S; Long. 114° 26.8W
Duration at site of greatest eclipse: 7 minutes, 25 seconds
Path of annularity: South Pacific Ocean; southern Chile and Argentina

For more detail on solar eclipse occurrences in the world, see *Fifty Year Canon of Solar Eclipses*, NASA Reference Publication 1178, by Dr. Fred Espenak.

Appendix II

Total lunar eclipses: 2001 to 2025

Below is a list of total lunar eclipses until 2025. The region where the Moon will be at the zenith at totality is noted. Generally, people located 7,500 to 8,000 kilometers east and west of this geographical location will witness all phases of the eclipse.

2001   January 9: Eastern Hemisphere
        Moon at zenith at totality: eastern Arabian Peninsula

2003   May 16: Western Hemisphere
        Moon at zenith at totality: southern Brazil

        November 9: western-Eastern Hemisphere
        Moon at zenith at totality: northwestern coast of Africa

2004   May 4: Western Hemisphere
        Moon at zenith at totality: Madagascar

        October 28: Western Hemisphere
        Moon at zenith at totality: Atlantic Ocean, north of South America

2007   March 3: Africa, Arabian Peninsula, northern Europe
        Moon at zenith at totality: Nigeria, Cameroon, Africa

        August 28: equatorial Pacific Ocean
        Moon at zenith at totality: French Polynesia

2008 February 21: Western Hemisphere
Moon at zenith at totality: Atlantic Ocean, north of South America

2010 December 21: western-Western Hemisphere
Moon at zenith at totality: off southwestern coast of the United States

2011 June 15: Eastern Hemisphere
Moon at zenith at totality: east of Madagascar

December 10: eastern-Eastern Hemisphere
Moon at zenith at totality: western North Pacific Ocean, Northern Mariana Islands

2014 April 15: Western Hemisphere
Moon at zenith at totality: equatorial Pacific Ocean, northeast of French Polynesia

October 8: western-Western Hemisphere
Moon at zenith at totality: mid-equatorial Pacific Ocean, south of Hawaii

2015 April 4: extreme western-Western Hemisphere
Moon at zenith at totality: equatorial Pacific Ocean, region of Palmyra Atoll

September 28: Western Hemisphere
Moon at zenith at totality: north coast of Brazil

2018 January 31: Eastern Hemisphere
Moon at zenith at totality: north equatorial Pacific Ocean, region of Marshall Islands

July 27: Eastern Hemisphere
Moon at zenith at totality: Indian Ocean, east of Madagascar

2019 January 21: Western Hemisphere
Moon at zenith at totality: Cuba

2021   May 26: western-Western Hemisphere
        Moon at zenith at totality: north South Pacific Ocean, region of Samoa

2022   May 16: Western Hemisphere
        Moon at zenith at totality: Bolivia, central South America

        November 8: western-Western Hemisphere
        Moon at zenith at totality: North Pacific Ocean, southwest of Hawaii

2025   March 14: Western Hemisphere
        Moon at zenith at totality: equatorial Pacific Ocean, west of Galápagos
        Islands

        September 7: Eastern Hemisphere
        Moon at zenith at totality: equatorial Indian Ocean, south of Sri Lanka

        For more detail on lunar eclipse occurrences in the world, see *Fifty Year
        Canon of Lunar Eclipses*, NASA Reference Publication 1216, by Dr. Fred
        Espenak.

Appendix III

General planet locations: 1999 to 2010

The tables below provide the approximate planet locations for the first day of every three months up to the year 2010. Each table lists the planet's name, whether it is an evening or morning object, and its host constellation. 'Morning' denotes planets that rise after midnight; 'evening' denotes planets that rise before midnight. 'Sun' indicates that the planet is too near the Sun to be seen.

As mentioned, planets like Mercury and Venus, and in some cases Mars, frequently change locations. To find a planet at a time not listed, you need to interpolate its position between the months given. For example, on January 1, 2004, Venus is an evening object in the constellation Capricornus. By April 1, 2004, it is still an evening object but now lies in Taurus. Thus, between January and April 2004, Venus moves through the region of sky between these two constellations – first into Aquarius, then Pisces.

Because of its rapid motion around the Sun, Mercury is the most difficult of the planets to keep track of. If you are intent on following Mercury back and forth between the morning and evening skies, I suggest you look up its position in a popular astronomy magazine or obtain monthly positions from an observer's almanac.

January 1, 1999

Mercury	morning	Ophiuchus
Venus	morning	Sagittarius
Mars	morning	Virgo
Jupiter	evening	Aquarius/Pisces

Saturn	evening	Pisces
Uranus	evening	Capricornus
Neptune	evening	Capricornus
Pluto	morning	Ophiuchus

April 1, 1999

Mercury	morning	Aquarius
Venus	evening	Aries
Mars	evening	Libra
Jupiter	Sun	Pisces
Saturn	evening	Pisces
Uranus	evening	Capricornus
Neptune	evening	Capricornus
Pluto	evening	Ophiuchus

July 1, 1999

Mercury	evening	Cancer
Venus	evening	Leo
Mars	evening	Virgo
Jupiter	morning	Aries
Saturn	morning	Aries
Uranus	evening	Capricornus
Neptune	evening	Capricornus
Pluto	evening	Ophiuchus

October 1, 1999

Mercury	evening	Virgo
Venus	morning	Leo
Mars	evening	Ophiuchus
Jupiter	evening	Aries
Saturn	evening	Aries
Uranus	evening	Capricornus
Neptune	evening	Capricornus
Pluto	evening	Ophiuchus

January 1, 2000

Mercury	morning	Sagittarius

Venus	morning	Scorpius
Mars	evening	Aquarius
Jupiter	evening	Pisces
Saturn	evening	Aries
Uranus	evening	Capricornus
Neptune	evening	Capricornus
Pluto	morning	Ophiuchus

April 1, 2000

Mercury	morning	Aquarius
Venus	morning	Aquarius
Mars	evening	Aries
Jupiter	evening	Aries
Saturn	evening	Aries
Uranus	morning	Capricornus
Neptune	morning	Capricornus
Pluto	evening	Ophiuchus

July 1, 2000

Mercury	Sun	Gemini
Venus	Sun	Gemini
Mars	Sun	Gemini
Jupiter	morning	Taurus
Saturn	morning	Taurus
Uranus	evening	Capricornus
Neptune	evening	Capricornus
Pluto	evening	Ophiuchus

October 1, 2000

Mercury	evening	Virgo
Venus	evening	Libra
Mars	morning	Leo
Jupiter	evening	Taurus
Saturn	evening	Taurus
Uranus	evening	Capricornus
Neptune	evening	Capricornus
Pluto	evening	Ophiuchus

January 1, 2001

Mercury	Sun	Sagittarius
Venus	evening	Aquarius
Mars	morning	Virgo
Jupiter	evening	Taurus
Saturn	evening	Taurus
Uranus	evening	Capricornus
Neptune	evening	Capricornus
Pluto	morning	Ophiuchus

April 1, 2001

Mercury	morning	Aquarius
Venus	morning	Pisces
Mars	morning	Ophiuchus
Jupiter	evening	Taurus
Saturn	evening	Taurus
Uranus	morning	Capricornus
Neptune	morning	Capricornus
Pluto	morning	Ophiuchus

July 1, 2001

Mercury	morning	Taurus
Venus	morning	Taurus
Mars	evening	Ophiuchus
Jupiter	morning	Taurus
Saturn	morning	Taurus
Uranus	evening	Capricornus
Neptune	evening	Capricornus
Pluto	evening	Ophiuchus

October 1, 2001

Mercury	evening	Virgo
Venus	morning	Leo
Mars	evening	Sagittarius
Jupiter	evening	Gemini
Saturn	evening	Taurus
Uranus	evening	Capricornus

| Neptune | evening | Capricornus |
| Pluto | evening | Ophiuchus |

January 1, 2002

Mercury	evening	Sagittarius
Venus	Sun	Sagittarius
Mars	evening	Aquarius
Jupiter	evening	Gemini
Saturn	evening	Taurus
Uranus	evening	Capricornus
Neptune	evening	Capricornus
Pluto	morning	Ophiuchus

April 1, 2002

Mercury	Sun	Pisces
Venus	evening	Aries
Mars	evening	Aries
Jupiter	evening	Gemini
Saturn	evening	Taurus
Uranus	morning	Aquarius
Neptune	morning	Capricornus
Pluto	evening	Ophiuchus

July 1, 2002

Mercury	morning	Taurus
Venus	evening	Leo
Mars	evening	Gemini
Jupiter	evening	Gemini
Saturn	morning	Taurus
Uranus	evening	Aquarius
Neptune	evening	Capricornus
Pluto	evening	Ophiuchus

October 1, 2002

| Mercury | Sun | Virgo |
| Venus | evening | Libra |

Mars	morning	Leo
Jupiter	morning	Cancer
Saturn	evening	Taurus
Uranus	evening	Capricornus
Neptune	evening	Capricornus
Pluto	evening	Ophiuchus

January 1, 2003

Mercury	evening	Sagittarius
Venus	morning	Libra
Mars	morning	Libra
Jupiter	evening	Cancer
Saturn	evening	Taurus
Uranus	evening	Capricornus
Neptune	evening	Capricornus
Pluto	morning	Ophiuchus

April 1, 2003

Mercury	evening	Pisces
Venus	morning	Aquarius
Mars	morning	Sagittarius
Jupiter	evening	Cancer
Saturn	evening	Taurus
Uranus	morning	Aquarius
Neptune	morning	Capricornus
Pluto	evening	Ophiuchus

July 1, 2003

Mercury	Sun	Gemini
Venus	morning	Taurus
Mars	evening	Aquarius
Jupiter	evening	Leo
Saturn	Sun	Gemini
Uranus	evening	Aquarius
Neptune	evening	Capricornus
Pluto	evening	Ophiuchus

October 1, 2003

Mercury	morning	Leo
Venus	evening	Virgo
Mars	evening	Aquarius
Jupiter	morning	Leo
Saturn	evening	Gemini
Uranus	evening	Aquarius
Neptune	evening	Capricornus
Pluto	evening	Ophiuchus

January 1, 2004

Mercury	morning	Sagittarius
Venus	evening	Capricornus
Mars	evening	Pisces
Jupiter	evening	Leo
Saturn	evening	Gemini
Uranus	evening	Aquarius
Neptune	evening	Capricornus
Pluto	morning	Serpens

April 1, 2004

Mercury	evening	Aries
Venus	evening	Taurus
Mars	evening	Taurus
Jupiter	evening	Leo
Saturn	evening	Gemini
Uranus	morning	Aquarius
Neptune	morning	Capricornus
Pluto	evening	Serpens

July 1, 2004

Mercury	evening	Gemini
Venus	morning	Taurus
Mars	evening	Cancer
Jupiter	evening	Leo
Saturn	Sun	Gemini
Uranus	evening	Aquarius

| Neptune | evening | Capricornus |
| Pluto | evening | Serpens |

October 1, 2004

Mercury	Sun	Virgo
Venus	morning	Leo
Mars	Sun	Virgo
Jupiter	Sun	Virgo
Saturn	morning	Gemini
Uranus	evening	Aquarius
Neptune	evening	Capricornus
Pluto	evening	Serpens

January 1, 2005

Mercury	morning	Ophiuchus
Venus	morning	Ophiuchus
Mars	morning	Scorpius
Jupiter	morning	Virgo
Saturn	evening	Gemini
Uranus	evening	Aquarius
Neptune	evening	Capricornus
Pluto	morning	Serpens

April 1, 2005

Mercury	Sun	Pisces
Venus	Sun	Pisces
Mars	morning	Capricornus
Jupiter	evening	Virgo
Saturn	evening	Gemini
Uranus	morning	Aquarius
Neptune	morning	Capricornus
Pluto	evening	Serpens

July 1, 2005

Mercury	evening	Cancer
Venus	evening	Cancer
Mars	morning	Pisces

Jupiter	evening	Virgo
Saturn	evening	Cancer
Uranus	evening	Aquarius
Neptune	evening	Capricornus
Pluto	evening	Serpens

October 1, 2005

Mercury	evening	Virgo
Venus	evening	Libra
Mars	evening	Taurus
Jupiter	evening	Virgo
Saturn	morning	Cancer
Uranus	evening	Aquarius
Neptune	evening	Capricornus
Pluto	evening	Serpens

January 1, 2006

Mercury	morning	Sagittarius
Venus	evening	Capricornus
Mars	evening	Aries
Jupiter	morning	Libra
Saturn	evening	Cancer
Uranus	evening	Aquarius
Neptune	evening	Capricornus
Pluto	morning	Serpens

April 1, 2006

Mercury	morning	Aquarius
Venus	morning	Capricornus
Mars	evening	Taurus
Jupiter	evening	Libra
Saturn	evening	Cancer
Uranus	morning	Aquarius
Neptune	morning	Capricornus
Pluto	morning	Serpens

July 1, 2006

Mercury	evening	Cancer
Venus	morning	Taurus
Mars	evening	Taurus
Jupiter	evening	Libra
Saturn	evening	Cancer
Uranus	evening	Aquarius
Neptune	evening	Capricornus
Pluto	evening	Serpens

October 1, 2006

Mercury	evening	Virgo
Venus	Sun	Virgo
Mars	Sun	Virgo
Jupiter	evening	Libra
Saturn	morning	Leo
Uranus	evening	Aquarius
Neptune	evening	Capricornus
Pluto	evening	Serpens

January 1, 2007

Mercury	Sun	Sagittarius
Venus	evening	Sagittarius
Mars	morning	Ophiuchus
Jupiter	morning	Ophiuchus
Saturn	evening	Leo
Uranus	evening	Aquarius
Neptune	evening	Capricornus
Pluto	morning	Serpens

April 1, 2007

Mercury	morning	Aquarius
Venus	evening	Aries
Mars	morning	Aries
Jupiter	morning	Ophiuchus
Saturn	evening	Leo
Uranus	morning	Aquarius

Neptune	morning	Capricornus
Pluto	morning	Serpens

July 1, 2007

Mercury	Sun	Gemini
Venus	evening	Leo
Mars	morning	Aries
Jupiter	evening	Ophiuchus
Saturn	evening	Leo
Uranus	evening	Aquarius
Neptune	evening	Capricornus
Pluto	evening	Serpens

October 1, 2007

Mercury	evening	Virgo
Venus	morning	Leo
Mars	evening	Gemini
Jupiter	evening	Ophiuchus
Saturn	morning	Leo
Uranus	evening	Aquarius
Neptune	evening	Capricornus
Pluto	evening	Sagittarius

January 1, 2008

Mercury	evening	Sagittarius/Capricornus
Venus	morning	Scorpius/Libra
Mars	evening	Taurus
Jupiter	morning	Sagittarius
Saturn	evening	Leo
Uranus	evening	Aquarius
Neptune	evening	Capricornus
Pluto	morning	Sagittarius

April 1, 2008

Mercury	Sun	Pisces
Venus	morning	Aquarius

Mars	evening	Gemini
Jupiter	morning	Sagittarius
Saturn	evening	Leo
Uranus	morning	Aquarius
Neptune	morning	Capricornus
Pluto	morning	Sagittarius

July 1, 2008

Mercury	morning	Taurus
Venus	Sun	Gemini
Mars	evening	Leo
Jupiter	evening	Sagittarius
Saturn	evening	Leo
Uranus	evening	Pisces
Neptune	evening	Capricornus
Pluto	evening	Sagittarius

October 1, 2008

Mercury	Sun	Virgo
Venus	evening	Libra
Mars	evening	Virgo
Jupiter	evening	Sagittarius
Saturn	morning	Leo
Uranus	evening	Aquarius
Neptune	evening	Capricornus
Pluto	evening	Sagittarius

January 1, 2009

Mercury	evening	Capricornus
Venus	evening	Aquarius
Mars	morning	Sagittarius
Jupiter	evening	Capricornus
Saturn	evening	Leo
Uranus	evening	Aquarius
Neptune	evening	Capricornus
Pluto	Sun	Sagittarius

April 1, 2009

Mercury	Sun	Pisces
Venus	morning	Pisces
Mars	morning	Aquarius
Jupiter	morning	Capricornus
Saturn	evening	Leo
Uranus	morning	Aquarius
Neptune	morning	Capricornus
Pluto	morning	Sagittarius

July 1, 2009

Mercury	morning	Taurus
Venus	morning	Taurus
Mars	morning	Aries
Jupiter	evening	Capricornus
Saturn	evening	Leo
Uranus	evening	Pisces
Neptune	evening	Capricornus
Pluto	evening	Sagittarius

October 1, 2009

Mercury	morning	Leo
Venus	morning	Leo
Mars	evening	Gemini
Jupiter	evening	Capricornus
Saturn	morning	Virgo
Uranus	evening	Pisces
Neptune	evening	Capricornus
Pluto	evening	Sagittarius

January 1, 2010

Mercury	Sun	Sagittarius
Venus	Sun	Sagittarius
Mars	evening	Leo
Jupiter	evening	Capricornus
Saturn	evening	Virgo
Uranus	evening	Aquarius

| Neptune | evening | Capricornus |
| Pluto | Sun | Sagittarius |

April 1, 2010

Mercury	evening	Pisces
Venus	evening	Aries
Mars	evening	Cancer
Jupiter	morning	Aquarius
Saturn	evening	Virgo
Uranus	morning	Pisces
Neptune	morning	Aquarius
Pluto	morning	Sagittarius

July 1, 2010

Mercury	Sun	Gemini
Venus	evening	Leo
Mars	evening	Leo
Jupiter	evening	Pisces
Saturn	evening	Virgo
Uranus	evening	Pisces
Neptune	evening	Aquarius
Pluto	evening	Sagittarius

October 1, 2010

Mercury	morning	Virgo
Venus	evening	Libra
Mars	evening	Libra
Jupiter	evening	Pisces
Saturn	Sun	Virgo
Uranus	evening	Pisces
Neptune	evening	Capricornus
Pluto	evening	Sagittarius

Appendix IV

Oppositions for Mars, Jupiter, and Saturn: 2000 to 2010

The table for each planet lists its dates of opposition with corresponding least distances from Earth.

Mars

2000 June 13	67 million km (42 million mi)
2003 August 28	56 million km (35 million mi)
2005 November 7	69 million km (43 million mi)
2007 December 24	88 million km (55 million mi)
2010 January 29	100 million km (62 million mi)

Jupiter

2000 November 28	606 million km (376 million mi)
2002 January 1	626 million km (390 million mi)
2003 February 2	647 million km (402 million mi)
2004 March 4	662 million km (411 million mi)
2005 April 3	667 million km (414 million mi)
2006 May 4	645 million km (400 million mi)
2007 June 6	635 million km (395 million mi)
2008 July 9	630 million km (390 million mi)
2009 August 14	590 million km (366 million mi)
2010 September 21	570 million km (355 million mi)

Saturn

2000 November 19	1.21 billion km (756 million mi)
2001 December 3	1.20 billion km (750 million mi)

2002 December 17   1.20 billion km (750 million mi)
2003 December 31   1.20 billion km (750 million mi)
2005 January 13   1.20 billion km (750 million mi)
2006 January 27   1.22 billion km (760 million mi)
2007 February 10   1.22 billion km (760 million mi)
2008 February 24   1.23 billion km (765 million mi)
2009 March 8   1.24 billion km (770 million mi)
2010 March 22   1.24 billion km (770 million mi)

Appendix V

The 20 brightest stars in the night sky

The 20 stars below comprise the brightest visible to the naked eye in the night sky in both hemispheres. The apparent magnitudes of the brightest stars are pretty well agreed upon in the astronomy literature. More problematic, however, are their distances. Note that some distances are accompanied by a second value in parentheses. This second value is merely an alternative estimate based on results that used slightly different parameters or observational and analytical methods to derive stellar distances. Of course, a single star cannot occupy space at two difference distances simultaneously, but it would be misleading for me to arbitrarily select a distance value that asserts unequivocally that Betelgeuse, say, is 540 light-years away, when other data, just as good, puts that value closer to 1,400. Hence, in cases of divergences, I've supplied at least two outside values and am more than content to let astronomers fight it out.

The interpretation of the various star names are usually derived from Latin or Arabic phrases. Aldebaran, for example, is from the Arabic *Al Dabaran*, the follower, and Betelgeuse is from *Ibt al Jauzah*, armpit of the central one. Like distances, however, there is a wide range of interpretations of these names. Those presented here are the ones more commonly used.

Star name	Interpretation	Bayer letter	Apparent magnitude	Distance in light-years
1 Sirius	The scorching one	α Canis Majoris	−1.46	8.65
2 Canopus	Chief pilot of the fleet of Menelaus	α Carinae	−0.72	75
3 Arcturus	Bear guard or bear watcher	α Boötis	−0.04	35
4 Rigel Kentaurus	The centaur's foot	α Centauri	0.00	4.39
5 Vega	The plunging one or falling eagle or vulture	α Lyrae	0.03	25
6 Capella	The little she-goat	α Aurigae	0.08	43
7 Rigel	Orion's left foot or leg	β Orionis	0.12	910 (1400)
8 Procyon	Before the dog	α Canis Minoris	0.38	11
9 Achernar	The end of the river	α Eridani	0.46	69
10 Betelgeuse	Armpit of the giant or central one	α Orionis	0.50	540 (1400)
11 Hadar	The ground	β Centauri	0.61	450 (320)
12 Altair	The flying eagle or vulture	α Aquilae	0.77	16
13 Aldebaran	The follower	α Tauri	0.85	60 (65)
14 Antares	The rival of Mars	α Scorpii	0.96	522 (440)
15 Spica	The ear of the corn	α Virginis	0.98	220 (260)
16 Pollux	The boxer	β Geminorum	1.14	35
17 Fomalhaut	The mouth of the southern fish	α Piscis Austrinus	1.16	22
18 Deneb	The hen's tail	α Cygni	1.25	1,500 (1,800)
19 Mimosa	Modern name for Beta Crucis; also Becrux	β Crucis	1.25	460 (580)
20 Regulus	The prince	α Leonis	1.35	69 (72)

Star name interpretations are based on 'Pronunciations, Derivations, and Meanings of a Selected List of Star Names' by George A. Davis, Jr., *Popular Astronomy*, January 1944; *Star Names: Their Lore and Meaning* by Richard Hinkley Allen (Dover Books, 1963); and *The New Patterns in the Sky* by Julius D. W. Staal (The McDonald and Woodward Publishing Company, 1988). Magnitudes and distances are based on a variety of available data, but most specifically the *Observer's Handbook of the Royal Astronomical Society of Canada*, 1998, and *Starlist 2000* by Richard Dibon-Smith (John Wiley and Sons, Inc., 1992).

Further Reading

Almanacs and annuals

Observer's Handbook, edited by Roy L. Bishop
  (Royal Astronomical Society of Canada)
The Astronomical Calendar, Guy Ottewell
  (Astronomical Workshop)
*The Handbook of the British Astronomical
  Association*, Gordon E. Taylor, Director
  (British Astronomical Association)

Astronomy enrichment

Daniel J. Boorstin, *The Discoverers* (Random
  House, Inc., 1983)
Sir James Frazer, *The Golden Bough: A Study in
  Magic and Religion* (Wordsworth Reference,
  1996)
Stephen J. Gould, *Questioning the Millennium*
  (Harmony Books, 1997)
George Johnson, *Fire in the Mind* (Alfred A.
  Knopf, Inc., 1995)
Martha Evans Martin and Donald Howard
  Menzel, *The Friendly Stars* (Dover
  Publications, 1966)
Chet Raymo, *The Soul of the Night* (Hungry
  Mind Press, 1992)
Carl Sagan, *The Demon-Haunted World*
  (Ballantine Books, 1996)

Archeoastronomy and history

Anthony F. Aveni, *Skywatchers of Ancient Mexico*
  (University of Texas Press, 1980)
Anthony F. Aveni, *Conversing with the Planets*
  (Times Books, 1992)
E. C. Krupp, *Beyond the Blue Horizon* (Oxford
  University Press, 1991)
Jean Guard Monroe and Ray A. Williamson,
  *They Dance in the Sky: Native American Star
  Myths* (Houghton Mifflin Company, 1987)
John North, *The Norton History of Astronomy
  and Cosmology* (W. W. Norton Company,
  1994)
Hugh Thurston, *Early Astronomy* (Springer-
  Verlag, 1994)

Celestial Tables, Catalogs, and Star Maps

Richard Dibon-Smith, *Starlist 2000* (John Wiley
  & Sons, 1992)
Milton D. Heifetz and Wil Tirion, *A Walk through
  the Heavens* (Cambridge University Press,
  1998)
Alan Hirshfeld and Roger W. Sinnott, *Sky
  Catalogue 2000.0* (Sky Publishing
  Corporation and Cambridge University Press,
  1985)
Jean Meeus, *Astronomical Tables of the Sun,
  Moon, and Planets* (Willmann-Bell, Inc., 1983)

Roger W. Sinnott, editor, *NGC 2000.0* (Cambridge University Press and Sky Publishing Corporation, 1988)

Roger W. Sinnott and Michael A. C. Perryman, *Millennium Star Atlas* (Sky Publishing Corporation and the European Space Agency, 1997)

General astronomy

George O. Abell, David Morrison, Sidney C. Wolff, *Exploration of the Universe*, sixth edition (Saunders College Publishing, 1993)

D. Scott Birney, *Observational Astronomy* (Cambridge University Press, 1991)

Valerie Illingworth, editor, *The Facts on File Dictionary of Astronomy* (Facts on File, 1994)

James B. Kaler, *The Ever-Changing Sky* (Cambridge University Press, 1996)

Guy Ottewell, *Astronomical Companion* (Universal Workshop, 1991)

Jay Pasachoff, *Journey Through the Universe* (Saunders College Publishing, 1994)

Stephen P. Maran, editor, *The Astronomy and Astrophysics Encyclopedia* (Van Nostrand Reinhold, 1992)

Michael Zeilik and Stephen A. Gregory, *Introductory Astronomy and Astrophysics*, fourth edition (Saunders College Publshing, 1998)

Observing guides

Robert Burnham, Alan Dyer, Robert A. Garfinkle, Martin George, Jeff Kanipe, David Levy, *Advanced Skywatching* (Time-Life Books, 1997)

Robert Burnham Jr., *Burnham's Celestial Handbook*, Vols. 1 – 3 (Dover Publications, 1978)

Neil Davis, *The Aurora Watcher's Handbook* (University of Alaska Press, 1992)

Terence Dickinson and Alan Dyer, *The Backyard Astronomer's Guide* (Camden House Publishing, 1991)

Stephen J. Edberg and David H. Levy, *Observing Comets, Asteroids, Meteors, and the Zodiacal Light* (Cambridge University Press, 1994)

Robert A. Garfinkle, *Star-Hopping: Your Visa to Viewing the Universe* (Cambridge University Press, 1997)

David Levy, *Skywatching* (Time-Life Books, 1994)

James Muirden, *Skywatcher's Handbook* (W.H. Freeman, 1993)

Jay Pasachoff, *Stars and Planets* (Peterson Field Guides, Houghton Mifflin Company, 1992)

Robin Scagell, *City Astronomy* (Sky Publishing, 1994)

Solar and Lunar eclipses

Fred Espenak, *Fifty Year Canon of Solar Eclipses: 1986 – 2035* (NASA Reference Publication 1178 Revised, July 1987)

Fred Espenak, *Fifty Year Canon of Lunar Eclipses: 1986 – 2035* (NASA Reference Publication 1216, March 1989)

Solar system

Fritz Heide and Frank Wlotzka, *Meteorites: Messengers from Space* (Springer-Verlag, 1994)

Rudolf Kippenhahn, *Discovering the Secrets of the Sun* (John Wiley & Sons, 1994)

Kim Long, *The Moon Book* (Johnson Books, 1988)

O. Richard Norton, *Rocks from Space* (Mountain Press Publishing Company, 1994)

Fred W. Price, *The Planet Observer's Handbook* (Cambridge University Press, 1998)

Paul Spudis, *The Once and Future Moon* (Smithsonian Institution Press, 1996)

Peter O. Taylor, *Observing the Sun* (Cambridge University Press, 1991)

Star lore and myths

Richard Hinckley Allen, *Star Names: Their Lore and Meaning* (Dover Publications, 1963)

George A. Davis, Jr., 'Pronunciations, Derivations, and Meanings of a Selected List of Star Names,' *Popular Astronomy*, January 1944, reprinted by Sky Publishing Corporation, 1971

Giuseppe Maria Sesti, *The Glorious Constellations: History and Mythology* (Abrams, Inc., Publishers, 1991)

Julius D. W. Staal, *The New Patterns in the Sky* (The McDonald and Woodward Publishing Company, 1988)

Jean-Pierre Verdet, *The Sky: Mystery, Magic, and Myth* (Harry N. Abrams, Inc., 1992)

Telescopes

Richard Berry, *Build Your Own Telescope* (Charles Scribner's Sons, 1985)

Sam Brown , *All About Telescopes* (Edmund Scientific Company)

Universe

Paul Davies, *The Last Three Minutes* (Basic Books, 1994)

Timothy Ferris, *The Whole Shebang* (Simon and Schuster, 1997)

Robert Jastrow, *God and the Astronomers* (W. W. Norton and Company, Inc., 1992)

Index